2742

Springer

Berlin
Heidelberg
New York
Barcelona
Hong Kong
London
Milan
Paris
Singapore
Tokyo

K. Beyreuther   Y. Christen   C. L. Masters (Eds.)

Neurodegenerative Disorders: Loss of Function Through Gain of Function

With 28 Figures and 5 Tables

Springer

Beyreuther, Konrad, Dr. rer. nat. Dr. med. h.c.
ZMBH
University of Heidelberg
Im Neuenheimer Feld 282
69120 Heidelberg, Germany
E-mail: beyreuther@zmbh.uni-heidelberg.de

Christen, Yves, Ph. D.
Fondation IPSEN
Pour la Recherche Thérapeutique
24, rue Erlanger
75781 Paris Cedex 16, France
E-mail: yves.christen@beaufour-ipsen.com

Masters, Colin L., M. D.
Department of Pathology and
The Mental Health Institute of Victoria
The University of Melbourne
Parkville, Victoria, 3052, Australia
E-mail: c.masters@pathology.unimelb.edu.au

ISSN 0945-6066
ISBN 3-540-41218-2   Springer-Verlag Berlin Heidelberg New York

Library of Congress Cataloging-in-Publication Data
Neurodegenerative disorders : loss of function through gain of function / K. Beyreuther,
Y. Christen, C.L. Masters (eds.)
p. cm. – (Research and perspectives in Alzheimer's disease)
Includes bibliographical references and index.
ISBN 3540412182 (hc. : alk. paper)
1. Nervous system – Degeneration. 2. Alzheimer's disease. 3. Amyloid
beta-protein – Pathophysiology. 4. Amyloid beta-protein precursor – Pathophysiology. I.
Beyreuther, K. (Konrad), 1941 – II. Christen, Yves. III. Masters, Colin L. IV. Series.

Springer-Verlag Berlin Heidelberg New York
a member of BertelsmannSpringer Science+Business Media GmbH
http://www.springer.de

© Springer-Verlag Berlin Heidelberg 2001
Printed in Germany

Production: PRO EDIT GmbH, 69126 Heidelberg, Germany
Typesetting: Mitterweger & Partner GmbH, Plankstadt

Printed on acid-free paper – SPIN: 10717992   27/3130göh 5 4 3 2 1 0

Preface

Fondation Ipsen sponsored a meeting in Paris in February 2000 on the emerging paradigm-shift in our understanding of the major degenerative diseases which affect the aging human brain. This book summarizes our deliberations on some of these major neurodegenerative diseases that are characterized by protein deposits, and that are due to the pathogenic gain of function of an otherwise normal neuronal protein.

For each of the major human neurodegenerative diseases covered in this book – the most prominent being Alzheimer's disease – experimental models are described, including cell culture systems and animal models which range from the round worm, *Caenorhabditis elegans*, the fruitfly, *Drosophila melanogaster*, to rodents. Remarkably, in the sporadic forms of these human diseases, only a minor change in the level of production or turn-over of the relevant proteins is sufficient to cause disease in late adult-hood. Neurodegeneration in Alzheimer's disease, for example, usually results in symptoms and signs in the seventh to eighth decades. In contrast, the development of protein deposits in transgenic mice over-expressing the corresponding disease gene parallels the genetic forms of the human diseases in regard to its manifestation occuring half-way through its normal life-span, i. e. about 50 years in humans (the so-called "presenium") and 9 to 12 months in the mouse. Nevertheless, these models have served to elucidate many of the pathways underlying the human disease processes, for instance clarifying the neuronal origin of parenchymal and perivascular amyloid in Alzheimer's disease and Creutzfeldt-Jakob disease. But there is much more to be learned. The animal and the cellular models will also allow the investigation of the intracellular and extracellular pathways involved in the causation of neurodegeneration.

The main message from this book is that the different protein aggregation processes may all be amenable to a small number of intervention steps based on a common theme of the modulation of production, turnover and deposition of the corresponding disease gene products. The next years will prove critical in evaluating the possibilities of rational therapeutic strategies towards regaining the loss of function through the amelioration of the abnormal gain of function.

January 2001

Konrad Beyreuther
Yves Christen
Colin Masters

Contents

List of Contributors

Abramowski, D.
Novartis Pharma, Ltd., 4002 Basel, Switzerland

Allinquant, B.
CNRS UMR 8542, Ecole Normale Supérieure, 46, rue d'Ulm, 75005 Paris, France

Annaert, W.
Neuronal Cell Biology and Gene Transfer Laboratory, Center for Human Genetics, Catholic University of Leuven and Flanders Interuniversitary Institute for Biotechnology, 3000 Leuven, Belgium

Beal, M.F.
Department of Neurology and Neuroscience, Weill Medical College of Cornell University and the New York Hospital – Cornell Medical Center, 525 East 68th Street, New York, NY 10021, USA
e-mail: fbeal@mail.med.cornell.edu

Beffert, U.
Department of Molecular Genetics, University of Texas Southwestern Medical Center, 53223 Harry Hines Blvd., Dallas, TX 75390–9046, USA

Beyreuther, K.
Center for Molecular Biology (ZMBH), The University of Heidelberg, 69120 Heidelberg, Germany
e-mail: beyreuther@zmbh.uni-heidelberg.de

Boncristiano, S.
Neuropathology, Institute of Pathology, University of Basel, 4003 Basel, Switzerland

Bondolfi, L.
Neuropathology, Institute of Pathology, University of Basel, 4003 Basel, Switzerland

Borchelt, D.R.
Department of Pathology, The Johns Hopkins University School of Medicine, 558 Ross Research Building, 720 Rutland Avenue, Baltimore, MD 21205–2196, USA

Budnik, V.
Department of Biology, University of Massachusetts, Amherst, Massachusetts
01003, USA

Calhoun, M.
Neuropathology, Institute of Pathology, University of Basel, 4003 Basel,
Switzerland
and Mount Sinai School of Medicine, New York, NY 10029, USA

Cleveland, D.W.
Ludwig Institute and Departments of Medicine and Neurosciences, University
of California at San Diego, 9500 Gilman Drive, La Jolla, CA 92093–0660, USA
e-mail: dcleveland@ucsd.edu

Craessaeerts, K.
Neuronal Cell Biology and Gene Transfer Laboratory, Center for Human
Genetics, Catholic University of Leuven and Flanders Interuniversitary
Institute for Biotechnology, 3000 Leuven, Belgium

Cupers, P.
Neuronal Cell Biology and Gene Transfer Laboratory, Center for Human
Genetics, Catholic University of Leuven and Flanders Interuniversitary
Institute for Biotechnology, 3000 Leuven, Belgium

De Strooper, B.
Neuronal Cell Biology and Gene Transfer Laboratory, Center for Human
Genetics, Catholic University of Leuven and Flanders Interuniversitary
Institute for Biotechnology, 3000 Leuven, Belgium

Deller, T.
Institute of Anatomy, University of Freiburg, 79001 Freiburg, Germany

Devys, D.
Institut de Génétique et Biologie Moléculaire et Cellulaire, INSERM/CNRS/Uni-
versité Louis Pasteur, BP 163, 67404 Illkirch Cedex, CU de Strasbourg, France

Goate, A.
Department of Psychiatry and Genetics, Washington University School
of Medicine, St. Louis, MO 63110, USA

Gotthardt, M.
Department of Molecular Genetics, University of Texas Southwestern Medical
Center, 53223 Harry Hines Blvd., Dallas, TX 75390–9046, USA

Hedge, R.S.
Laboratory of Cellular Oncology, National Institutes of Health, Bethesda, MD
20892, USA

Herreman, A.
Neuronal Cell Biology and Gene Transfer Laboratory, Center for Human
Genetics, Catholic University of Leuven and Flanders Interuniversitary
Institute for Biotechnology, 3000 Leuven, Belgium

Herz, J.
Department of Molecular Genetics, University of Texas Southwestern Medical
Center, 53223 Harry Hines Blvd., Dallas, TX 75390–9046, USA
e-mail: herz@utsw.swmed.edu

Herzig, M.
Neuropathology, Institute of Pathology, University of Basel, 4003 Basel,
Switzerland

Hiesberger, T.
Department of Molecular Genetics, University of Texas Southwestern Medical
Center, 53223 Harry Hines Blvd., Dallas, TX 75390–9046, USA

Huppert, S.
Department of Molecular Biology and Pharmacology, Washington University,
St. Louis, MO 63110, USA

Jucker, M.
Neuropathology, Institute of Pathology, University of Basel, 4003 Basel,
Switzerland
e-mail: mjucker@uhbs.ch

Kopan, R.
Department of Molecular Biology and Pharmacology, Washington University,
St. Louis, MO 63110, USA

Lee, M.K.
Department of Pathology, The Johns Hopkins University School of Medicine,
558 Ross Research Building, 720 Rutland Avenue, Baltimore, MD 21205–2196,
USA

Lingappa, V.R.
Departments of Physiology and Medicine, University of California,
San Francisco, CA 94143-0444, USA

Lunkes, A.
Institut de Génétique et Biologie Moléculaire et Cellulaire, INSERM/CNRS/Uni-
versité Louis Pasteur, BP 163, 67404 Illkirch Cedex, CU de Strasbourg, France

Mainguy, G.
CNRS UMR 8542, Ecole Normale Supérieure, 46, rue d'Ulm, 75005 Paris,
France

Mandel, J.-L.
Institut de Génétique et Biologie Moléculaire et Cellulaire, INSERM/CNRS/Université Louis Pasteur, BP 163, 67404 Illkirch Cedex, CU de Strasbourg, France

Markowska, A.L.
Department of Psychology, The Johns Hopkins University School of Medicine, 558 Ross Research Building, 720 Rutland Avenue, Baltimore, MD 21205–2196, USA

Martin, L.J.
Department of Pathology, The Johns Hopkins University School of Medicine, 558 Ross Research Building, 720 Rutland Avenue, Baltimore, MD 21205–2196, USA

Masters, C.L.
Department of Pathology, The University of Melbourne, Parkville, Victoria, 3052 and The Mental Research Institute of Victoria, Australia
e-mail: c.masters@pathology.unimelb.edu.au

Mumm, J.S.
Department of Molecular Biology and Pharmacology, Washington University, St. Louis, MO 63110, USA

Packard, M.
Department of Biology, University of Massachusetts, Amherst, MA 01003, USA

Pfeifer, M.
Neuropathology, Institute of Pathology, University of Basel, 4003 Basel, Switzerland

Phinney, A.
Neuropathology, Institute of Pathology, University of Basel, 4003 Basel, Switzerland

Price, D.L.
Division of Neuropathology, The Johns Hopkins University School of Medicine, 558 Ross Research Building, 720 Rutland Avenue, Baltimore, MD 21205–2196, USA
e-mail: priced@jhmi.edu

Probst, A.
Neuropathology, Institute of Pathology, University of Basel, 4003 Basel, Switzerland

Prochiantz, A.
CNRS UMR 8542, Ecole Normale Supérieure, 46, rue d'Ulm, 75005 Paris, France
e-mail: prochian@wotan.ens.f

Ray, W.J.
Department of Psychiatry and Genetics, Washington University School
of Medicine, St. Louis, MO 63110, USA

Rothstein, J.
Department of Neurology, The Johns Hopkins University School of Medicine,
558 Ross Research Building, 720 Rutland Avenue, Baltimore, MD 21205–2196,
USA

Saxena, M.T.
Department of Molecular Biology and Pharmacology, Washington University,
St. Louis, MO 63110, USA

Schroeter, E.H.
Department of Molecular Biology and Pharmacology, Washington University,
St. Louis, MO 63110, USA

Serneels, L.
Neuronal Cell Biology and Gene Transfer Laboratory, Center for Human
Genetics, Catholic University of Leuven and Flanders Interuniversitary
Institute for Biotechnology, 3000 Leuven, Belgium

Sisodia, S.S.
Department of Pharmacological and Physiological Sciences,
The University of Chicago, Chicago, IL 60637, USA

Sommer, B.
Novartis Pharma, Ltd., 4002 Basel, Switzerland

Stalder, M.
Neuropathology, Institute of Pathology, University of Basel, 4003 Basel,
Switzerland

Staufenbiel, M.
Novartis Pharma, Ltd., 4002 Basel, Switzerland

Sturchler-Pierrat, C.
Novartis Pharma, Ltd., 4002 Basel, Switzerland

Thinakaran, G.
Department of Pharmacological and Physiological Sciences,
The University of Chicago, Chicago, IL 60637, USA

Tolnay, M.
Neuropathology, Institute of Pathology, University of Basel, 4003 Basel,
Switzerland

Torroja, L.
Department of Biology and Center for Complex Systems, Brandeis University,
Waltham, MA 02454, USA

Trottier, Y.
Institut de Génétique et Biologie Moléculaire et Cellulaire, INSERM/CNRS/Université Louis Pasteur, BP 163, 67404 Illkirch Cedex, CU de Strasbourg, France

White, K.
Department of Biology and Center for Complex Systems, Brandeis University, Waltham, MA 02454, USA

Wiederhold, K.-H.
Novartis Pharma, Ltd., 4002 Basel, Switzerland

Williamson, T.L.
Trophos, Parc de Luminy, Batiment CCIMP – Case 922, 13288 Marseille Cedex 9, France

Winkler, D.
Neuropathology, Institute of Pathology, University of Basel, 4003 Basel, Switzerland

Wong, P.C.
Department of Pathology, The Johns Hopkins University School of Medicine, 558 Ross Research Building, 720 Rutland Avenue, Baltimore, MD 21205–2196, USA

Yvert, G.
Institut de Génétique et Biologie Moléculaire et Cellulaire, INSERM/CNRS/Université Louis Pasteur, BP 163, 67404 Illkirch Cedex, CU de Strasbourg, France

The Natural History of Alzheimer's Disease: Minding the Gaps in Understanding the Mechanisms of Neurodegeneration

C. L. Masters and K. Beyreuther

Summary

Alzheimer's disease (AD) remains a formidable challenge despite advances in our understanding of many of the molecular events which surround its development. Studies over the past 20 years have focused on the role of Aβ amyloid in this disorder, with the aim of developing rational therapeutic strategies which can modify or prevent the disease process. This approach has been vindicated by the general acceptance of the "Aβ amyloid hypothesis," which, among other concepts, has delivered the therapeutic targets of the β- and γ-secretases of the amyloid protein precursor (APP) which form the basis of the biogenesis of Aβ amyloid, the presenilins which are also intimate participants in the generation of Aβ, and other interacting proteins such as ApoE for which genetic evidence exists of linkage to the AD process. Despite this wealth of new knowledge, there remain many large gaps in our understanding of the pathogenesis of AD. These gaps include:
1) a full description of quantifiable levels of Aβ in relation to the progression of AD, and whether Aβ levels in biological fluids (plasma, CSF) can be used as reliable biological markers of the disease process;
2) whether currently available transgenic mouse models of AD sufficiently replicate the human disease, and whether the failure of these mice to develop neurofibrillary tangles is a major impediment;
3) the physical forms of Aβ which are pathologically relevant to the neurodegenerative process;
4) the intracellular compartments in which these etiologically relevant forms of Aβ are produced;
5) how extracellular forms of Aβ are processed and cleared from the brain, and whether either the intracellular or extracellular locations of Aβ can account for the peculiar topographic vulnerability of the brain in AD.

Minding these gaps should lead to a more complete account of the natural history of AD and should also elucidate new therapeutic strategies.

Introduction

Alzheimer's disease (AD) remains a formidable challenge despite dramatic advances in our knowledge of many of the molecular events which surround its occurrence (for recent reviews, see Haass and De Strooper 1999; Koo et al. 1999;

Research and Perspectives in Alzheimer's Diseases
Beyreuther/Christen/Masters (Eds.)
Neurodegenerative Disorders
© Springer-Verlag Berlin Heidelberg 2001

Masters and Beyreuther 1998; Roher et al. 1999; Selkoe 1999; Wilson et al. 1999). Although the peculiar neurodegenerative changes described by Alzheimer have now been known for more than a century, it was not until 1927 that Divry first drew attention to the presence of amyloid within the "senile" plaque. Thus began the "Aβ amyloid hypothesis of AD." The modern era of AD research began nearly 30 years ago with the first attempts at the biochemical characterisation of the

Table 1. Milestones towards formulating and testing the Aβ amyloid hypothesis of Alzheimer's disease

1892–1907	Lesions associated with Alzheimer's neurodegeneration described and categorized (Alzheimer 1907; Blocq and Marinesco 1892; Redlich 1898)
1920–40	Amyloid plaques and perivascular deposits described and proposed to adversely affect cerebral function (Divry 1927; Scholz 1938)
1970–86	Isolation (Nikaido et al. 1971), amino acid composition (Allsop et al. 1983; Roher et al. 1986; Selkoe et al. 1986) and N-terminal sequences of perivascular (Glenner and Wong 1984) and parenchymal plaque Aβ amyloid (Masters et al. 1985)
1987–88	Amyloid precursor protein (APP) gene cloned, located on chromosome 21 and Aβ recognized as putative proteolytic product (Kang et al. 1987; Tanzi et al. 1987, 1988; Goldgaber et al. 1987; Robakis et al. 1987; Kitaguchi et al. 1988; Ponte et al. 1988)
1989–99	Toxicity of Aβ in cell culture (Whitson et al. 1989; Yankner et al. 1989), mediated by oxidative processes (Behl et al. 1992, 1994; Dyrks et al. 1992; Butterfield et al. 1994) and metals (Bush et al. 1994; Schubert and Chevion 1995; Huang et al. 1999; Cherny et al. 1999)
1990–92	Mutations in proximity to Aβ secretase sites cause AD (Levy et al. 1990; Chartier-Harlin et al. 1991; Goate et al. 1991; Murrell et al. 1991; Citron et al. 1992; Hendriks et al. 1992) and confirm centrality of Aβ hypothesis
1992–93	Aβ and p3 identified as definite proteolytic products from APP (Seubert et al. 1992, 1993; Esch et al. 1990; Golde et al. 1992; Haass et al. 1992, 1993; Shoji et al. 1992; Busciglio et al. 1993)
1992–93	ApoE identified as an Aβ-interacting protein (Wisniewski and Frangione 1992; Strittmatter et al. 1993) and its alleles identified as genetic risk factors in sporadic forms of AD (Corder et al. 1993)
1991–97	Transgenic mouse models of Aβ deposition (Quon et al. 1991; Borchelt et al. 1997; Sturchler-Pierrat et al. 1997; Games et al. 1995; Hsiao et al. 1996)
1995–2000	Presenilins recognized as major components in the amyloidogenic pathway (Levy-Lahad 1995; Borchelt et al. 1996; Sherrington et al. 1995; Rogaev et al. 1995; Scheuner et al. 1996), probably acting as γ-secretases (Ray et al. 1999; Capell et al. 2000; Kimberly et al. 2000; Jacobsen et al 1999; Octave et al. 2000; De Strooper et al. 1998, 1999; Wolfe et al. 1999a, b; Steiner et al. 1999)
1998–99	α-secretases identified (Buxbaum et al. 1998; Lammich et al. 1999)
1999–2000	Major Aβ degradative and clearance pathways identified (Iwata et al. 2000; Qiu et al. 1999; Vekrellis et al. 2000) and improved clearance of Aβ from the brain by immunisation with Aβ (Schenk et al. 1999)
1999–2000	β-secretases identified (Vassar et al. 1999; Yan et al. 1999; Sinha et al. 1999; Lin et al. 2000; Hussain et al. 1999)
1995–2000	First rational anti-amyloid therapeutic strategies (Higaki et al. 1995) and commencement of early phase human trials (Bristol Myers Squibb, Elan Pharmaceuticals, Prana Biotechnology)

amyloid protein. Since then, several milestones have been attained which have focused on the role of Aβ amyloid in this disorder (Table 1). These efforts have been based on the understanding that modulation of Aβ accumulation in the brain might represent the most rational approach towards therapeutic intervention in this disease. This approach has been vindicated by the general acceptance of the "Aβ amyloid hypothesis of AD," which was strengthened substantially by the realisation that mutations in proximity to the Aβ domain of the amyloid precursor protein (APP) actually cause aggressive forms of early-onset AD. Among other concepts, pursuit of the Aβ hypothesis has now delivered the therapeutic targets of the β-secretases (Grüninger-Leitch et al. 2000; Hussain et al. 1999; Lin et al. 2000; Sinha et al. 1999; Vassar et al. 1999; Yan et al. 1999) and the γ-secretases/presenilins (Capell et al. 2000; De Strooper et al. 1998, 1999; Georgakopoulos et al. 1999; Jacobsen et al. 1999; Katayama et al. 1999; Kimberley et al. 2000; Niwa et al. 1999; Octave et al. 2000; Palacino et al. 2000; Pradier et al. 1999; Ray et al. 1999; Shirotani et al. 2000; Steiner et al. 1999; Weidemann et al. 1997; Wolfe et al. 1999a, b). These secretases act directly on APP, which then forms the central pathway in the molecular pathogenesis of AD (Fig. 1). Other interacting proteins, such as ApoE, have been identified which form a network of factors which drive the central pathway towards neurodegeneration, with consequent neuronal dysfunction and widespread neurologic and cognitive impairment. The

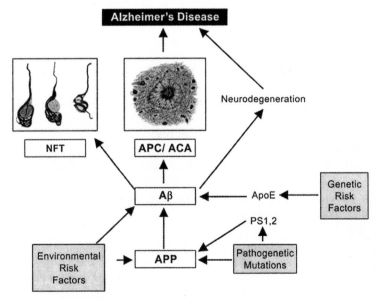

Fig. 1. A schematic outline of the pathway leading to Alzheimer's disease (AD). The processing of APP into Aβ is modulated by both environmental factors (yet to be identified) and pathogenic mutations in the PS, APP and possibly other genes. Genetic risk factors identified to date (ApoE, α2M) may operate at the level of Aβ turnover in the brain. Amyloid plaques are the visible result of this pathway but do not of themselves cause the neurodegeneration of AD. The mechanism of neurofibrillary tangle (NFT) formation remains enigmatic

major test of the Aβ hypothesis is now about to commence, with early clinical trials of therapeutic strategies aimed at the Aβ amyloidogenic pathway commencing in 2000 (Table 1).

Despite this wealth of new knowledge, there remain many large gaps in our understanding of the pathogenesis of AD as currently based on the Aβ amyloid hypothesis. These gaps include:

1) a full description of quantifiable levels of Aβ in relation to the progression of AD, and whether Aβ levels in biological fluids (plasma, CSF) can be used as reliable biological markers of the disease process;
2) whether currently available transgenic mouse models of AD sufficiently replicate the human disease, and whether the failure of these mice to develop neurofibrillary tangles is a major impediment;
3) the physical forms of Aβ which are pathologically relevant to the Alzheimer neurodegenerative process;
4) the intracellular compartments in which these pathologically relevant forms of Aβ are produced;
5) how extracellular forms of Aβ are processed and cleared form the brain, and whether either the intracellular or extracellular locations of Aβ can account for the peculiar topographic vulnerability of the brain in AD.

Minding these gaps, which form the basis of this review, should lead to a more complete account of the natural history of AD, as distinguished from the normal aging process, and could also elucidate further rational therapeutic strategies.

Levels of Aβ in Human Brain, CSF and Blood

If Aβ amyloid were both a marker and a cause of AD, then quantitation of absolute levels of Aβ in the brain should disclose a relationship with the natural history of the disease. For a variety of reasons which are now becoming clear, this relationship has been difficult to establish, yet it does exist.

In the normal aging human brain, low levels of soluble and insoluble forms of Aβ are present (McLean et al. 1999; Näslund et al. 2000) and increase markedly in the end-stages of disease (see Table 2). In a large postmortem cross-sectional study of nursing home residents, a good correlation was found between total Aβ brain levels and the degree of cognitive impairment which encompassed the full clinical spectrum from mild to severe dementia (Näslund et al. 2000). While the larger proportion of insoluble Aβ as plaques correlates poorly with the degree of neuropathologic damage in the end-stage AD brain (McLean et al. 1999), a good correlation has been identified with the soluble Aβ levels (Lue et al. 1999; McLean et al. 1999; Wang et al. 1999). The Western blot techniques used to identify and assay this soluble Aβ pool have distinct advantages over the more commonly used two-site enzyme immunoassays (see also Enya et al. 1999; Funato et al. 1999) and serve to emphasise the continued need for assay development. A further important distinction is to be made in assays of $A\beta_{40}$ vs $A\beta_{42}$, where these

Table 2. Mean levels of Aβ (μg/g) in human and transgenic mouse brain[a]

	Soluble Aβ	Insoluble Aβ	Proportion soluble (%)
Human			
Controls (n=18)	<0.1	1.9±2.45	–
AD (n=18)	0.3±0.25**	20.6±11.05	1.4
Transgenic mouse APP Tg2576			
15 months	N/A	350	–
23 months	5	770	0.6
C100			
12 months	0.12±0.01	0.27±0.04	44

[a] From McLean et al. 1999 and Li and Masters (unpublished observations).

species have differing cellular origins and biological properties (see below). Nevertheless, confirmation of the soluble pool of Aβ as the major determinant of neurotoxic activity will have major therapeutic implications. From these preliminary studies, the picture of different brain pools of Aβ is emerging (see Fig. 2). The interrelationships between these pools and those present in the CSF and blood can now be studied in more detail.

Attempts at measuring Aβ levels in CSF have produced discordant results (Andreasen et al. 1999; Hulstaert et al. 1999; Ida et al. 1996; Jensen et al. 1999; Mehta et al. 2000; Samuels et al. 1999; Schröder et al. 1997; Tapiola et al. 2000), with some investigators finding elevated levels in early stages and decreased levels in later stages. While it appears that CSF Aβ is increased in those forms of

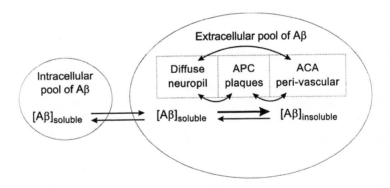

Fig. 2. The putative different pools of Aβ. The intracellular biogenesis of Aβ provides the initial pool, some of which may also exist in a relatively insoluble state. Release of Aβ by as yet unknown mechanisms into the extracellular space provides the source of Aβ which eventually precipitates as plaques and perivascular deposits. The levels of soluble Aβ, which correlate with the severity of the Alzheimer neurodegenerative process, may derive from either (or both) the intracellular and extracellular compartments. These levels will also depend on the capacity of the extracellular compartment to act as a "sink" and on those mechanisms which normally remove Aβ from the brain

familial AD and Down's syndrome associated with an overproduction of Aβ, overall it appears that CSF levels of Aβ are depressed in sporadic AD, suggesting that there is a generalised failure of Aβ clearance from the brain parenchyma in AD. However, the relative half-lives and contributions of the choroid plexus and plasma Aβ to the CSF pools have not yet been explored and factored into these CSF estimations.

Our studies of the origins of plasma Aβ (predominantly $A\beta_{40}$ species) have disclosed that the platelet is a major source (Li et al. 1994, 1998, 1999a; Whyte et al. 1997). Initial studies of Aβ levels in plasma have also disclosed wide variability and difficulties working at the limits of detectability. More recent studies have shown encouraging correlations of elevated plasma $A\beta_{42}$ with AD (Mayeux et al. 1999) and the realisation of the effect of Aβ complexed to other carrier molecules, which needs to be taken into account (Kuo et al. 1999; Matsubara et al. 1999). Thus it will be important to fully understand the kinetics of $A\beta_{40, 42}$ turn-over in plasma in relation to the use of these assays as markers of AD or as indices of efficacy of therapeutic intervention.

Replication of AD in Transgenic Mice

A major problem exists with the existing transgenic mouse models of AD in that, to date, all have failed to develop florid neurofibrillary tangles (and neuritic changes) and the behavioural phenotypes are mild (see Emilien et al. 2000 for review and Calhoun et al. 1999; Chapman et al. 1999; Li et al. 1999b; McGowan et al. 1999; Sberna et al. 1998). Nevertheless, the models currently available have proven of immense value in efforts to understand the mechanisms of Aβ amyloid deposition and turnover (Calhoun et al. 1999), the role of Aβ in causing neurode-generation (Chapman et al. 1999; Holcomb et al. 1999; Hsia et al. 1999; McGowan et al. 1999) and the effects of overexpression of intracellular forms of Aβ (Chui et al. 1999; Li et al. 1999b). The results to date are consistent with our concept of soluble forms of non-plaque Aβ being the major determinant of neurodegeneration (section 1 above), but the failure to generate neurofibrillary tangles, possibly linked to the failure to develop a robust behavioral phenotype, raises interesting questions: have the rodent brains been sufficiently challenged by the overexpression of Aβ; are rodents inherently resistant to tau polymerisation within a one- to two-year time period; does the presence of wild-type rodent Aβ inhibit tangle formation?

The actual levels of transgenic mouse brain Aβ far exceed those determined in human AD (see Table 2); thus it seems unlikely that the challenge is inadequate. Since it appears that rodent tau is capable of polymerising into paired helical filaments, at least in vitro (Kampers et al. 1999), then it seems more likely that there is a fundamental link missing in the Aβ-tau concept. Further novel transgenic models are therefore required to resolve this issue (e.g., use of null background mice, crosses of Aβ overexpressing mice with oxidative stress sensitive mice, and constructs aimed at high level intracellular $A\beta_{42}$ production).

The existing models also provide excellent substrates to address mechanisms of clearance of Aβ from the brain and the validation of therapeutic strategies aimed either at interfering with Aβ production or the solubilization and clearance of Aβ.

The Pathologically Relevant Biophysical Forms of Aβ

Our focus on the soluble forms of Aβ (Hilbich et al. 1991; McLean et al. 1999) has been sharpened by findings on methods to solubilize the detergent-resistant forms of human aggregated Aβ (Cherny et al. 1999). Similar principles may underlie the aggregation of α-synuclein in diffuse Lewy body disease and multiple system atrophy (Conway et al. 2000).

Most previous and more recent biophysical studies on the soluble forms of Aβ have concentrated on synthetic $A\beta_{40}$ (Hartley et al. 1999; Harper et al. 1999; Kowalewski and Holtzman 1999; Walsh et al. 1999). These studies need to be confirmed and complemented by an analysis of the forms of Aβ which actually constitute the soluble pool in the human AD brain. Particular aspects relating to N- and C-terminal truncations (Pillot et al. 1999a; Tjernberg et al. 1999), distinguishing secreted $A\beta_{40}$ from intracellular $A\beta_{42}$, post-translational modifications (such as dityrosine bridge formation and other oxidative modifications; Galeazzi et al. 1999), and interacting molecules (Bohrmann et al. 1999; Mason et al. 1999; Matsuzaki and Horikiri 1999; Shao et al. 1999; Yang et al. 1999a) can then be studied from a pathologically relevant basis.

The first step in this process is to prepare larger quantities of human AD and normal brain Aβ soluble species, and to compare these to those extracted from Tg mouse models. In addition to these biochemical studies, it will be important to confirm the "toxicity" of these pathological Aβ soluble species in cell-based assays, which could ultimately lead to assay systems for drug development (Huang et al. 1999). The toxicity of synthetic Aβ has been studied by many investigators over the last several years, yet the species actually responsible for the neurodegeneration in the AD brain, and particularly whether first- or second-order kinetics operate for the presumptive mode of toxicity of these Aβ species, remains to be determined.

Intracellular Sites of Aβ Production

Some of the fundamental questions concerning the Aβ amyloid hypothesis concern the mechanisms of translocation of the Aβ molecule from a transmembrane-embedded compartment to its accumulation within the cerebral extracellular space, the cellular compartments of origin of Aβ (and whether $A\beta_{40}$ and $A\beta_{42}$ have different origins), and the neuron versus other cells of origin in relation to the brain-restricted nature of AD.

Our studies to date have emphasised the possible micro-environment (including metal ions, extracellular matrix components) in which APP and Aβ occur in both intracellular and extracellular compartments (Bush et al. 1993; Fuller et al. 1995; Multhaupt et al. 1996, 1998; Sberna et al. 1997; Williamson et al. 1996) and earlier studies have been carried out on neuronal versus non-neuronal cells in their respective mechanisms of processing APP (Moir et al. 1998). More recent studies from our laboratories (Culvenor et al. 1997; Fuller et al. 1995; Hartmann et al. 1997; Tienari et al. 1997) and by others (e.g., Annaert et al. 1999; Gouras et al. 2000; Greenfield et al. 1999; Le Blanc and Goodyer 1999; Soriano et al. 1999; Yang et al. 1999b) have identified distinct intracellular compartments in which $A\beta_{40}$ and $A\beta_{42}$ can be generated, and have shown that the orientation of $A\beta$ in the membrane or that the physical nature of the membrane (e.g., its thickness) may play a role in $A\beta_{40, 42}$ production (Lichtenthaler et al. 1999). There is an emerging consensus that $A\beta_{42}$ may be generated in a pre-Golgi/endoplasmic reticulum compartment and normally retained and degraded intracellularly, whereas $A\beta_{40}$ is generated in an endosomal/plasma membrane compartment with a greater tendency for release into the extracellular space (Skovronsky et al. 2000).

These findings become very pertinent when we consider the mechanisms of $A\beta$ production in neurons, and the specialisations associated with axonal targeting of APP through the juxtamembranous domains of exon 15 (Hartmann et al. 1996; Sandbrink et al. 1994a, b; Tienari et al. 1996a, b). Moreover, the interactions between APP and presenilins (Evin et al. 2000; Weidemann et al. 1997) can now be re-interpreted in the light of vesicular trafficking of APP into the axonal/synaptic compartments (Beher et al. 1999; Clarris et al. 1994, 1995; Dumanchin et al. 1999; Georgakopoulos et al. 1999; Sekijima et al. 1999; Storey et al. 1996a, b, 1999), and transport to the cell surface of both full-length APP (Culvenor et al. 1995; Storey et al. 1996a, b; 1999) and presenilin (Ray et al. 1999; Schwarzman et al. 1999). Internalisation of these cell surface APP/$A\beta$ species into endosomal/lysosomal compartments may also play a major role in determining the site of $A\beta$ production, turnover and release (Evin et al. 1995; Li et al. 1999c; Octave et al. 1999; Peraus et al. 1997; Perez et al. 1999; Parkin et al. 1999). The quantities of $A\beta$ available for release will also be governed by the cell's own degradative machinery (Christie et al. 1999; Mok et al. 1997; Weidemann et al. 1999) and its quality control mechanisms for dealing with hydrophobic segments derived from transmembrane domains of molecules such as APP (Wickner et al. 1999).

Thus, if one accepts the prevailing view that the more toxic $A\beta_{42}$ species are generated in a pre-Golgi/ER compartment, it will be necessary to examine the consequences of this situation in the cerebral environment, in which polarised neurons have several options for degradation/release of small hydrophobic peptide fragments. In terms of the pathogenesis of neurodegeneration, we are particularly interested in determing whether $A\beta_{40}$ or $A\beta_{42}$ ever translocates into the cytoplasm (preliminary data indicate that this occurs; Gouras et al. 2000; Hartmann et al. 1997,) and whether either of these species is released from neurons in a presynaptic/postsynaptic environment (i.e., either into the synaptic cleft or into a space immediately surrounding the synapse).

Extracellular Cerebral Aβ: Uptake, Degradation and Clearance

Clearance from the brain of Aβ$_{40/42}$ released from neurons is probably the least studied and understood aspect of AD, which is surprising since the morphologic patterns of distribution of aggregated Aβ in the laminae of the cortical neuropil and around the medium-to-small arterioles speak strongly in favour of the importance of this basic pathway. We and others have begun the process of unravelling this pathway (Culvenor et al. 1999; Chung et al. 1999; Ida et al. 1996; Iwata et al. 2000; Morelli et al. 1999; Vekrellis et al. 2000; Yamin et al. 1999). Although microglia, because of their intimate association with the amyloid plaque, are logical effectors for the degradation of aggregated Aβ, the studies of Chung et al. (1999) suggest that the microglia might act through a process of solubilisation and mobilisation of insoluble Aβ, rather than actual degradation of Aβ. This finding would be consistent with a growing body of evidence that the only generally accepted genetic risk factor for AD, apolipoprotein E, also acts in a fashion which directly affects the balance between deposition and clearance of Aβ from the brain (Aleshkov et al. 1999; Bales et al. 1999; Holtzman et al. 1999; Moir et al. 1999; Pillot et al. 1999b; Yang et al. 1999c). Similar data are emerging for other molecules for which the genetic linkage data are less firm (Blacker et al. 1998; Qiu et al. 1999; Scharnagl et al. 1999). Other proteins which bind Aβ directly may subserve primary protective roles in the brain and other peripheral tissues (Bohrmann et al. 1999) through their actions of maintaining Aβ in a diffusible state, thereby allowing it to be efficiently cleared through the natural processes of bulk flow towards the Virchow-Robin spaces and the transependymal CSF compartment.

A very surprising result which appears to be consistent with the above concepts of clearance has been recently described by Schenk et al. (1999), in which immunisation of transgenic APP mice with human Aβ$_{42}$ [at either younger (1½ month) or older (11 month) ages] prevented the deposition of Aβ as plaques and the associated pathologic changes. It is even probable that pre-formed plaques could be removed by shifting the equilibrium towards the clearance mechanism. Although many aspects of this approach require further clarification, it does suggest that knowledge of the mechanisms of clearance of Aβ through immunological mediation will permit a better understanding of the turnover of Aβ and assist in the process of differentiating the toxic effects of Aβ in either intracellular or extracellular cerebral compartments. Methods for labelling Aβ are beginning to be developed (e.g., Poduslo et al. 1999) and could be used in a variety of quantitative in vivo clearance assays.

Conclusion

The recognition that AD is but one example of a growing number of diseases associated with the aggregation, deposition, and toxicity of a variety of proteins (Table 3) is now permitting a more rational approach towards therapeutic inter-

Table 3. Neurodegenerative disease associated with the toxic gain of function of aggregated proteins.

Disease/condition	Protein
Alzheimer's disease Congophilic angiopathy	Proteolytically derived Aβ amyloid from the amyloid precursor protein (APP)
Creutzfeldt-Jakob disease Kuru Gerstmann-Sträussler-Scheinker syndrome Fatal insomnia	Conversion of the prion protein (PrP^c) into an abnormal conformer (PrP^{sc})
Parkinson's disease Diffuse Lewy body disease Multiple system atrophy	α-Synuclein
Motor neuron disease (amyotrophic lateral sclerosis)	Cu Zu superoxide dismutase (SOD-1)
Fronto-temporal degeneration with Parkinsonism (FTDP) Pick's disease Progressive supranuclear palsy (Steele-Richardson-Olszewski syndrome)	Tau microtuble-associated protein
Huntington's disease Spinocerebellar ataxias Kennedy's syndrome	Polyglutamine expansions (triplet repeat diseases)
Worster-Drought syndrome (dementia with spastic paraparesis)	Novel (A$_{Bri}$) amyloidogenic protein
Cerebral autosomal dominant adult onset arteriopathy (CADASIL)	Notch-3 receptor extracellular domain

vention. The coming years will see "proof-of-principle" therapies emerge, first in critical transgenic animal models and then hopefully in humans. There has never been a better time for young investigators to join the endeavour to describe the natural histories of these illnesses and to help "mind the gaps" in our knowledge of their underlying pathogenic mechanisms.

References

Aleshkov SB, Li X, Lavrentiadou SN, Zannis VI (1999) Contribution of cysteine 158, the glycosylation site threonine 194, the amino- and carboxy-terminal domains of apolipoprotein E in the binding to amyloid peptide β (1–40). Biochemistry 38: 8918–8925

Allsop D, Landon M, Kidd M (1983) The isolation and amino acid composition of senile plaque core protein. Brain Res 259: 348–352

Alzheimer A (1907) Über eine eigenartige Erkrankung der Hirnrinde. Allgemeine Zeitschrift für Psychiatrie und Psychisch-Gerichtliche Medizin 64: 146–148

Andreasen N, Hesse C, Davidsson P, Minthon L, Wallin A, Winblad B, Vanderstichele H, Vanmechelen E, Blennow K (1999) Cerebrospinal fluid β-amyloid (1–42) in Alzheimer's disease: differences between early- and late-onset Alzheimer's disease and stability during the course of disease. Arch Neurol 56: 673–680

Annaert G, Levesque L, Craessaerts K, Dierinck I, Snellings G, Westaway D, George-Hyslop PS, Cordell B, Fraser P, De Strooper B (1999) Presenilin 1 controls γ-secretase processing of amyloid precursor protein in pre-Golgi compartments of hippocampal neurons. J Cell Biol 147: 277–294

Bales KR, Verina T, Cummins DJ, Du Y, Dodel RC, Saura J, Fishman CE, DeLong CA, Piccardo P, Petegnief V, Ghetti B, Paul SM (1999) Apolipoprotein E is essential for amyloid deposition in the APP (V717F) transgenic mouse model of Alzheimer's disease. Proc Natl Acad Sci USA 96: 15233–15238

Beher D, Elle C, Underwood J, Davis JB, Ward R, Karran E, Masters CL, Beyreuther K, Multhaup G (1999) Proteolytic fragments of Alzheimer's disease-associated presenilin 1 are present in synaptic organelles and growth cone membranes of rat brain. J Neurochem 72: 1564–1573

Behl C, Davis J, Cole GM, Schubert D (1992) Vitamin E protects nerve cells from amyloid β protein toxicity. Biochem Biophys Res Commun 186: 944–950

Behl C, Davis JB, Lesley R, Schubert D (1994) Hydrogen peroxide mediates amyloid β protein toxicity. Cell 77: 817–827

Blacker D, Wilcox MA, Laird NM, Rodes L, Horvath SM, Go RCP, Perry R, Watson B, Bassett SS, McInnis MG, Albert MS, Hyman BT, Tanzi RE (1998) Alpha-2 macroglobulin is genetically associated with Alzheimer disease. Nature Genet 19: 357–360

Blocq P, Marinesco G (1892) Sur les lésions et la pathogénése de l'épilepsie dite essentielle. Semaine Méd 12: 445

Bohrmann B, Tjernberg L, Kuner P, Poli S, Levet-Trafit B, Naslund J, Richards G, Huber W, Dobeli H, Nordstedt C (1999) Endogenous proteins controlling amyloid β-peptide polymerization. Possible implications for β-amyloid formation in the central nervous system and in peripheral tissues. J Biol Chem 274: 15990–15995

Borchelt DR, Thinakaran G, Eckman CB, Lee MK, Davenport F, Ratovitsky T, Prada CM, Kim G, Seekins S, Yager D, Slunt HH, Wang R, Seeger M, Levey AI, Gandy SE, Copeland NG, Jenkins NA, Price DL, Younkin SG, Sisodia SS (1996) Familial Alzheimer's disease-linked presenilin 1 variants elevate Aβ1–42/1–40 ratio in vitro and in vivo. Neuron 17: 1005–1013

Borchelt DR, Ratovitski T, van Lare J, Lee MK, Gonzales V, Jenkins NA, Copeland NG, Price DL, Sisodia SS (1997) Accelerated amyloid deposition in the brains of transgenic mice coexpressing mutant presenilin 1 and amyloid precursor proteins. Neuron 19: 939–945

Busciglio J, Gabuzda DH, Matsudaira P, Yankner BA (1993) Generation of β-amyloid in the secretory pathway in neuronal and nonneuronal cells. Proc Natl Acad Sci USA 90: 2092–2096

Bush AI, Multhaup G, Moir RD, Williamson TG, Small DH, Rumble B, Pollwein P, Beyreuther K, Masters CL (1993) A novel zinc (II) binding site modulates the function of the beta A4 amyloid protein precursor of Alzheimer's disease. J Biol Chem 268: 16109–16112

Bush AI, Pettingell WH, Multhaup G, d Paradis M, Vonsattel JP, Gusella JF, Beyreuther K, Masters CL, Tanzi RE (1994) Rapid induction of Alzheimer Aβ amyloid formation by zinc. Science 265: 1464–1467

Butterfield DA, Hensley K, Harris M, Mattson M, Carney J (1994) β-Amyloid peptide free radical fragments initiate synaptosomal lipoperoxidation in a sequence-specific fashion: implications to Alzheimer's disease. Biochem Biophys Res Commun 200: 710–715

Buxbaum JD, Liu KN, Luo Y, Slack JL, Stocking KL, Peschon JJ, Johnson RS, Castner BJ, Cerretti DP, Black RA (1998) Evidence that tumor necrosis factor α converting enzyme is involved in regulated alpha-secretase cleavage of the Alzheimer amyloid protein precursor. J Biol Chem 273: 27765–27767

Calhoun ME, Burgermeister P, Phinney AL, Stalder M, Tolnay M, Wiederhold KH, Abramowski D, Sturchler-Pierrat C, Sommer B, Staufenbiel M, Jucker M (1999) Neuronal overexpression of mutant amyloid precursor protein results in prominent deposition of cerebrovascular amyloid. Proc Natl Acad Sci USA 96: 14088–14093

Capell A, Steiner H, Romig H, Keck S, Baader M, Grim MG, Baumeister R, Haass C (2000) Presenilin-1 differentially facilitates endoproteolysis of the β-amyloid precursor protein and Notch. Nature Cell Biol 2: 205–211

Chapman PF, White GL, Jones WM, Cooper-Blacketer D, Marshall VJ, Irizarry M, Younkin L, Good MA, Bliss TV, Hyman BT, Younkin SG, Hsiao KK (1999) Impaired synaptic plasticity and learning in aged amyloid precursor protein transgenic mice. Nature Neurosci 2: 271–276

Chartier-Harlin MC, Crawford F, Houlden H, Warren A, Hughes D, Fidani L, Goate A, Rossor M, Roques P, Hardy J, Mullan M (1991) Early-onset Alzheimer's disease caused by mutations at codon 717 of the β-amyloid precursor protein gene. Nature 353: 844–846

Cherny RA, Legg JT, McLean CA, Fairlie DP, Huang X, Atwood CS, Beyreuther K, Tanzi RE, Masters CL, Bush AI (1999) Aqueous dissolution of Alzheimer's disease Aβ amyloid deposits by biometal depletion. J Biol Chem 274: 23223–23228

Christie G, Markwell RE, Gray CW, Smith L, Godfrey F, Mansfield F, Wadsworth H, King R, McLaughlin M, Cooper DG, Ward RV, Howlett DR, Hartmann T, Lichtenthaler SF, Beyreuther K, Underwood J, Gribble SK, Cappai R, Masters CL, Tamaoka A, Gardner RL, Rivett AJ, Karran EH, Allsop D (1999) Alzheimer's disease: correlation of the suppression of β-amyloid peptide secretion from cultured cells with inhibition of the chymotrypsin-like activity of the proteasome. J Neurochem 73: 195–204

Chui DH, Tanahashi H, Ozawa K, Ikeda S, Checler F, Ueda O, Suzuki H, Araki W, Inoue H, Shirotani K, Takahasi K, Gallyas F, Tabira T (1999) Transgenic mice with Alzheimer presenilin 1 mutations show accelerated neurodegeneration without amyloid plaque formation. Nat Med 5: 560–564

Chung H, Brazil MI, Soe TT, Maxfield FR (1999) Uptake, degradation, and release of fibrillar and soluble forms of Alzheimer's amyloid β-peptide by microglial cells. J Biol Chem 274: 32301–32308

Citron M, Oltersdorf T, Haass C, McConlogue L, Hung AY, Seubert P, Vigo-Pelfrey C, Lieberburg I, Selkoe DJ (1992) Mutation of the β-amyloid precursor protein in familial Alzheimer's disease increases β-protein production. Nature 360: 672–674

Clarris HJ, Nurcombe V, Small DH, Beyreuther K, Masters CL (1994) Secretion of nerve growth factor from septum stimulates neurite outgrowth and release of the amyloid protein precursor of Alzheimer's disease from hippocampal explants. J Neurosci Res 38: 248–258

Clarris HJ, Key B, Beyreuther K, Masters CL, Small DH (1995) Expression of the amyloid protein precursor of Alzheimer's disease in the developing rat olfactory system. Brain Res Dev Brain Res 88: 87–95

Conway KA, Lee SJ, Rochet JC, Ding TT, Williamson RE, Lansbury PT Jr (2000) Acceleration of oligomerization, not fibrillization, is a shared property of both α-synuclein mutations linked to early-onset Parkinson's disease: implications for pathogenesis and therapy. Proc Natl Acad Sci USA 97: 571–576

Corder EH, Saunders AM, Strittmatter WJ, Schmechel DE, Gaskell PC, Small GW, Roses AD, Haines JL, Pericak-Vance MA (1993) Gene dose of apolipoprotein E type 4 allele and the risk of Alzheimer's disease in late onset families. Science 261: 921–923

Culvenor JG, Friedhuber A, Fuller SJ, Beyreuther K, Masters CL (1995) Expression of the amyloid precursor protein of Alzheimer's disease on the surface of transfected HeLa. Exp Cell Res 220: 474–481

Culvenor JG, Maher F, Evin G, Malchiodi-Albedi F, Cappai R, Underwood JR, Davis JB, Karran EH, Roberts GW, Beyreuther K, Masters CL (1997) Alzheimer's disease-associated presenilin 1 in neuronal cells: evidence for localization to the endoplasmic reticulum-Golgi intermediate compartment. J Neurosci Res 49: 719–731

Culvenor JG, McLean CA, Cutt S, Campbell BC, Maher F, Jakala P, Hartmann T, Beyreuther K, Masters CL, Li QX (1999) Non-Aβ component of Alzheimer's disease amyloid (NAC) revisited: NAC and α-synuclein are not associated with Aβ amyloid. Am J Pathol 155: 1173–1181

De Strooper B, Saftig P, Craessaerts K, Vanderstichele H, Guhde G, Annaert W, Vonfigura K, Vanleuven F (1998) Deficiency of presenilin-1 inhibits the normal cleavage of amyloid precursor protein. Nature 391: 387–390

De Strooper B, Annaert W, Cupers P, Saftig P, Craessaerts K, Mumm JS, Schroeter EH, Schrijvers V, Wolfe MS, Ray WJ, Goate A, Kopan R (1999) A presenilin-1-dependent γ-secretase-like protease mediates release of Notch intracellular domain. Nature 398: 518–522

Divry P (1927) Étude histochimique des plaques séniles. J Belge Neurol Psych 27: 643–657

Dyrks T, Dyrks E, Hartmann T, Masters C, Beyreuther K (1992) Amyloidogenicity of βA4 and βA4-bearing amyloid protein precursor fragments by metal-catalyzed oxidation. J Biol Chem 267: 18210–18217

Dumanchin C, Czech C, Campion D, Cuif MH, Poyot T, Martin C, Charbonnier F, Goud B, Pradier L, Frebourg T (1999) Presenilins interact with Rab11, a small GTPase involved in the regulation of vesicular transport. Human Mol Genet 8: 1263–1269

Emilién G, Maloteaux JM, Beyreuther K, Masters CL (2000) Alzheimer disease: mouse models pave the way for therapeutic opportunities. Arch Neurol 57: 176–181

Enya M, Morishima-Kawashima M, Yoshimura M, Shinkai Y, Kusui K, Khan K, Games D, Schenk D, Sugihara S, Yamaguchi H, Ihara Y (1999) Appearance of sodium dodecyl sulfate-stable amyloid β-protein (Aβ) dimer in the cortex during aging. Am J Pathol 154: 271–279

Esch FS, Keim PS, Beattie EC, Blacher RW, Culwell AR, Oltersdorf T, McClure D, Ward PJ (1990) Cleavage of amyloid β peptide during constitutive processing of its precursor. Science 248: 1122–1124

Evin G, Cappai R, Li QX, Culvenor JG, Small DH, Beyreuther K, Masters CL (1995) Candidate γ-secretases in the generation of the carboxyl terminus of the Alzheimer's disease βA4 amyloid: possible involvement of cathepsin D. Biochemistry 34: 14185–14192

Evin G, Le Brocque D, Culvenor JG, Cappai D, Weidemann A, Beyreuther K, Masters CL, Cappai R (2000) Presenilin 1 expression in yeast lowers secretion of the amyloid precursor protein. NeuroReport 11: 405–408

Fuller SJ, Storey E, Li QX, Smith AI, Beyreuther K, Masters CL (1995) Intracellular production of βA4 amyloid of Alzheimer's disease: modulation by phosphoramidon and lack of coupling to the secretion of the amyloid precursor protein. Biochemistry 34: 8091–8098

Funato H, Enya M, Yoshimura M, Morishima-Kawashima M, Ihara Y (1999) Presence of sodium dodecyl sulfate-stable amyloid β-protein dimers in the hippocampus CA1 not exhibiting neurofibrillary tangle formation. Am J Pathol 155: 23–28

Galeazzi L, Ronchi P, Franceschi C, Giunta S (1999) In vitro peroxidase oxidation induces stable dimers of β-amyloid (1–42) through dityrosine bridge formation. Amyloid: Int J Exp Clin Invest 6: 7–13

Games D, Adams D, Alessandrini R, Barbour R, Berthelette P, Blackwell C, Carr T, Clemens J, Donaldson T, Gillespie F, Guido T, Hagopian S, Johnson-Wood K, Khan K, Lee M, Leibowitz P, Lieberburg I, Little S, Masliah E, McConlogue L, Montoya-Zavala M, Mucke L, Paganini L, Penniman E, Power M, Schenk D, Seubert P, Snyder B, Soriano F, Tan H, Vitale J, Wadsworth S, Wolozin B, Zhao J (1995) Alzheimer-type neuropathology in transgenic mice overexpressing V717F β-amyloid precursor protein. Nature 373: 523–527

Georgakopoulos A, Marambaud P, Efthimiopoulos S, Shioi J, Cui W, Li HC, Schutte M, Gordon R, Holstein GR, Martinelli G, Mehta P, Friedrich VL Jr, Robakis NK (1999) Presenilin-1 forms complexes with the cadherin/catenin cell-cell adhesion system and is recruited to intercellular and synaptic contacts. Mol Cell 4: 893–902

Glenner GG, Wong CW (1984) Alzheimer's disease: initial report of the purification and characterization of a novel cerebrovascular amyloid protein Biochem Biophys Res Commun 120: 885–890

Goate A, Chartier-Harlin MC, Mullan M, Brown J, Crawford F, Fidani L, Giuffra L, Haynes A, Irving N, James L, Mant R, Newton P, Rooke K, Roques P, Talbot C, Pericak-Vance M, Roses A, Williamson R, Rossor M, Owen M, Hardy J (1991) Segregation of a missense mutation in the amyloid precursor protein gene with familial Alzheimer's disease. Nature 349: 704–706

Goldgaber D, Lerman MI, McBride OW, Saffiotti U, Gajdusek DC (1987) Characterization and chromosomal localization of a cDNA encoding brain amyloid of Alzheimer's disease. Science 235: 877–880

Golde TE, Estus S, Younkin LH, Selkoe DJ, Younkin SG (1992) Processing of the amyloid protein precursor to potentially amyloidogenic derivatives. Science 255: 728–730

Gouras GK, Tsai J, Naslund J, Vincent B, Edgar M, Checler F, Greenfield JP, Haroutunian V, Buxbaum JD, Xu H, Greengard P, Relkin NR (2000) Intraneuronal Aβ42 accumulation in human brain. Am J Pathol 156: 15–20

Greenfield JP, Tsai J, Gouras GK, Hai B, Thinakaran G, Checler F, Sisodia SS, Greengard P, Xu H (1999) Endoplasmic reticulum and trans-Golgi network generate distinct populations of Alzheimer β-amyloid peptides. Proc Natl Acad Sci USA 96: 742–747

Grüninger-Leitch F, Berndt P, Langen H, Nelboeck P, Dobeli H (2000) Identification of β-secretase-like activity using a mass spectrometry-based assay system. Nature Biotechnol 18: 66–70

Haass C, De Strooper B (1999) Presenilins in Alzheimer's disease-proteolysis hold the key. Science 286: 916–919

Haass C, Schlossmacher MG, Hung AY, Vigo-Pelfrey C, Mellon A, Ostaszewski BL, Lieberburg I, Koo EH, Schenk D, Teplow DB (1992) Amyloid β-peptide is produced by cultured cells during normal metabolism. Nature 359: 322–325

Haass C, Hung AY, Schlossmacher MG, Teplow DB, Selkoe DJ (1993) β-Amyloid peptide and a 3-kDa fragment are derived by distinct cellular mechanisms. J Biol Chem 268: 3021–3024

Harper JD, Wong SS, Lieber CM, Lansbury PT Jr (1999) Assembly of Aβ amyloid protofibrils: an *in vitro* model for a possible early event in Alzheinmer's disease. Biochemistry 38: 8972–8980

Hartley DM, Walsh DM, Ye CP, Diehl T, Vasquez S, Vassilev PM, Teplow DB, Selkoe DJ (1999) Proto-fibrillar intermediates of amyloid β-protein induce acute electrophysiological changes and progressive neurotoxicity in cortical neurons. J Neurosci 19: 8876–8884

Hartmann T, Bergsdorf C, Sandbrink R, Tienari PJ, Multhaup G, Ida N, Bieger S, Dyrks T, Weidemann A, Masters CL, Beyreuther K (1996) Alzheimer's disease βA4 protein release and amyloid precursor protein sorting are regulated by alternative splicing. J Biol Chem 271: 13208–13214

Hartmann T, Bieger SC, Bruhl B, Tienari PJ, Ida N, Allsop D, Roberts GW, Masters CL, Dotti CG, Unsicker K, Beyreuther K (1997) Distinct sites of intracellular production for Alzheimer's disease Aβ-40/42 amyloid peptides. Nature Med 3: 1016–1020

Hendriks L, van Duijn CM, Cras P, Cruts M, Van Hul W, van Harskamp F, Warren A, McInnis MG, Antonarakis SE, Martin JJ, Hofman A, van Broeckhoven C (1992) Presenile dementia and cerebral haemorrhage linked to a mutation at codon 692 of the β-amyloid precursor protein gene. Nature Genet 1: 218–221

Higaki J, Quon D, Zhong Z, Cordell B (1995) Inhibition of β-amyloid formation identifies proteolytic precursors and subcellular site of catabolism. Neuron 14: 651–659

Hilbich C, Kisters-Woike B, Reed J, Masters CL, Beyreuther K (1991) Human and rodent sequence analogs of Alzheimer's amyloid βA4 share similar properties and can be solubilized in buffers of pH 7.4. Eur J Biochem 201: 61–69

Holcomb LA, Gordon MN, Jantzen P, Hsiao K, Duff K, Morgan D (1999) Behavioral changes in transgenic mice expressing both amyloid precursor protein and presenilin-1 mutations: lack of association with amyloid deposits. Behav Genet 29: 177–185

Holtzman DM, Bales KR, Wu S, Bhat P, Parsadanian M, Fagan AM, Chang LK, Sun Y, Paul SM (1999) Expression of human apolipoprotein E reduces amyloid-β deposition in a mouse model of Alzheimer's disease. J Clin Invest 103: R15–R21

Hsia AY, Masliah E, McConlogue L, Yu GQ, Tatsuno G, Hu K, Kholodenko D, Malenka RC, Nicoll RA, Mucke L (1999) Plaque-independent disruption of neural circuits in Alzheimer's disease mouse models. Proc Natl Acad Sci USA 96: 3228–3233

Hsiao K, Chapman P, Nilsen S, Eckman C, Harigaya Y, Younkin S, Yang F, Cole G (1996) Correlative memory deficits, Aβ elevation, and amyloid plaques in transgenic mice. Science 274: 99–102

Huang X, Cuajungco MP, Atwood CS, Hartshorn MA, Tyndall JD, Hanson GR, Stokes KC, Leopold M, Multhaup G, Goldstein LE, Scarpa RC, Saunders AJ, Lim J, Moir RD, Glabe C, Bowden EF, Masters CL, Fairlie DP, Tanzi RE, Bush A (1999) Cu (II) potentiation of Alzheimer Aβ neurotoxicity. Correlation with cell-free hydrogen peroxide production and metal reduction. J Biol Chem 274: 37111–37116

Hulstaert F, Blennow K, Ivanoiu A, Schoonderwaldt HC, Riemenschneider M, De Deyn PP (1999) Improved discrimination of AD patients using β-amyloid (1–42) and tau levels in CSF. Neurology 52: 1555–1562

Hussain I, Powell D, Howlett DR, Tew DG, Meek TD, Chapman C, Gloger IS, Murphy KE, Southan CD, Ryan DM, Smith TS, Simmons DL, Walsh FS, Dingwall C, Christie G (1999) Identification of a novel aspartic protease (Asp 2) as β-secretase. Mol Cell Neurosci 4: 419–427

Ida N, Hartmann T, Pantel J, Schroder J, Zerfass R, Forstl H, Sandbrink R, Masters CL, Beyreuther K (1996) Analysis of heterogeneous βA4 peptides in human cerebrospinal fluid and blood by a newly developed sensitive Western blot assay. J Biol Chem 271: 22908–22914

Iwata N, Tsubuki S, Takaki Y, Watanabe K, Sekiguchi M, Hosoki E, Kawashima-Morishima M, Lee HJ, Hama E, Sekine-Aizawa Y, Saido TC (2000) Identification of the major Aβ1–42-degrading catabolic pathway in brain parenchyma: suppression leads to biochemical and pathological deposition. Nature Med 6: 143–150

Jacobsen H, Reinhardt D, Brockhaus M, Bur D, Kocyba C, Kurt H, Grim MG, Baumeister R, Loetscher H (1999) The influence of endoproteolytic processing of familial Alzheimer's disease presenilin 2 on βA42 amyloid peptide formation. J Biol Chem 274: 35233–35239

Jensen M, Schroder J, Blomberg M, Engvall B, Pantel J, Ida N, Basun H, Wahlund LO, Werle E, Jauss M, Beyreuther K, Lannfelt L, Hartmann T (1999) Cerebrospinal fluid Aβ42 is increased early in sporadic Alzheimer's disease and declines with disease progression. Ann Neurol 45: 504–511

Kampers T, Pangalos M, Geerts H, Wiech H, Mandelkow E (1999) Assembly of paired helical filaments from mouse tau: implications for the neurofibrillary pathology in transgenic mouse models for Alzheimer's disease. FEBS Lett 451: 39–44

Kang J, Lemaire HG, Unterbeck A, Salbaum JM, Masters CL, Grzeschik KH, Multhaup G, Beyreuther K, Müller-Hill B (1987) The precursor of Alzheimer's disease amyloid A4 protein resembles a cell-surface receptor. Nature 325: 733–736

Katayama T, Imaizumi K, Sato N, Miyoshi K, Kudo T, Hitomi J, Morihara T, Yoneda T, Gomi F, Mori Y, Nakano Y, Takeda J, Tsuda T, Itoyama Y, Murayama O, Takashima A, St George-Hyslop P, Takeda M, Tohyama M (1999) Presenilin-1 mutations downregulate the signalling pathway of the unfolded-protein response. Nature Cell Biol 1: 479–485

Kimberly WT, Xia W, Rahmati T, Wolfe MS, Selkoe DJ (2000) The transmembrane aspartates in presenilin 1 and 2 are obligatory for γ-secretase activity and amyloid β-protein generation. J Biol Chem 275: 3173–3178

Kitaguchi N, Takahashi Y, Tokushima Y, Shiojiri S, Ito H (1988) Novel precursor of Alzheimer's disease amyloid protein shows protease inhibitory activity. Nature 331: 530–532

Koo EH, Lansbury PT Jr, Kelly JW (1999) Amyloid diseases: abnormal protein aggregation in neurodegeneration Proc Natl Acad Sci USA 96: 9989–9990

Kowalewski T, Holtzman DM (1999) In situ atomic force microscopy study of Alzheimer's β-amyloid peptide on different substrates: new insights into mechanism of β-sheet formation. Proc Natl Acad Sci USA 96: 3688–3693

Kuo YM, Emmerling MR, Lampert HC, Hempelman SR, Kokjohn TA, Woods AS, Cotter RJ, Roher AE (1999) High levels of circulating Aβ42 are sequestered by plasma proteins in Alzheimer's disease. Biochem Biophys Res Commun 257: 787–791

Lammich S, Kojro E, Postina R, Gilbert S, Pfeiffer R, Jasionowski M, Haass C, Fahrenholz F (1999) Constitutive and regulated α-secretase cleavage of Alzheimer's amyloid precursor protein by a disintegrin metalloprotease. Proc Natl Acad Sci USA 96: 3922–3927

LeBlanc AC, Goodyer CG (1999) Role of endoplasmic reticulum, endosomal-lysosomal compartments, and microtubules in amyloid precursor protein metabolism of human neurons. J Neurochem 72: 1832–1842

Levy E, Carman MD, Fernandez-Madrid IJ, Power MD, Lieberburg I, van Duinen SG, Bots GT, Luyendijk W, Frangione B (1990) Mutation of the Alzheimer's disease amyloid gene in hereditary cerebral hemorrhage, Dutch type. Science 248: 1124–1126

Levy-Lahad E, Wasco W, Poorkaj P, Romano DM, Oshima J, Pettingell WH, Yu CE, Jondro PD, Schmidt SD, Wang K, Crowley AC, Fu Y, Guenette SY, Galas D, Nemens E, Wijsman EM, Bird TD, Schellenberg GD, Tanzi RE (1995) Candidate gene for the chromosome 1 familial Alzheimer's disease locus. Science 269: 973–977

Li QX, Berndt MC, Bush AI, Rumble B, Mackenzie I, Friedhuber A, Beyreuther K, Masters CL (1994) Membrane-associated forms of the βA4 amyloid protein precursor of Alzheimer's disease in human platelet an brain: surface expression on the activated human platelet. Blood 84: 133–142

Li QX, Whyte S, Tanner JE, Evin G, Beyreuther K, Masters CL (1998) Secretion of Alzheimer's disease Aβ amyloid peptide by activated human platelets. Lab Invest 78: 461–469

Li QX, Fuller SJ, Beyreuther K, Masters CL (1999a) The amyloid precursor protein of Alzheimer's disease in human brain and blood. J Leukoc Biol 66: 567–574

Li QX, Maynard C, Cappai R, McLean CA, Cherny RA, Lynch T, Culvenor JG, Trevaskis J, Tanner JE, Bailey KA, Czech C, Bush AI, Beyreuther K, Masters CL (1999b) Intracellular accumulation of detergent-soluble amyloidogenic Aβ fragment of Alzheimer's disease precursor protein in the hippocampus of aged transgenic mice. J Neurochem 72: 2479–2487

Li Y, Xu C, Schubert D (1999c) The up-regulation of endosomal-lysosomal components in amyloid β-resistant cells. J Neurochem 73: 1477–1482

Lichtenthaler SF, Wang R, Grimm H, Uljon SN, Masters CL, Beyreuther K (1999) Mechanism of the cleavage specificity of Alzheimer's disease γ-secretase identified by phenylalanine-scanning mutagenesis of the transmembrane domain of the amyloid precursor protein. Proc Natl Acad Sci USA 96: 3053–3058

Lin X, Koelsch G, Wu S, Downs D, Dashti A, Tang J (2000) Human aspartic protease memapsin 2 cleaves the β-secretase site of β-amyloid precursor protein. Proc Natl Acad Sci USA 97: 1456–1460

Lue LF, Kuo YM, Roher AE, Brachova L, Shen Y, Sue L, Beach T, Kurth JH, Rydel RE, Rogers J (1999) Soluble amyloid β peptide concentration as a predictor of synaptic change in Alzheimer's disease. Am J Pathol 155: 853–862

Masliah E, McConlogue L, Yu GQ, Tatsuno G, Hu K, Kholodenko D, Malenka RC, Nicoll RA, Mucke L (1999) Plaque-independent disruption of neural circuits in Alzheimer's disease mouse models. Proc Natl Acad Sci USA 96: 3228–3233

Mason RP, Jacob RF, Walter MF, Mason PE, Avdulov NA, Chochina SV, Igbavboa U, Wood WG (1999) Distribution and fluidizing action of soluble and aggregated amyloid β-peptide in rat synaptic plasma membranes. J Biol Chem 274: 18801–18807

Masters CL, Beyreuther K (1998) Alzheimer's disease. Brit Med J 316: 446–448

Masters CL, Simms G, Weinman NA, Multhaup G, McDonald BL, Beyreuther K (1985) Amyloid plaque core protein in Alzheimer disease and Down syndrome. Proc Natl Acad Sci USA 82: 4245–4249

Matsubara E, Ghiso J, Frangione B, Amari M, Tomidokoro Y, Ikeda Y, Harigaya Y, Okamoto K, Shoji M (1999) Lipoprotein-free amyloidogenic peptides in plasma are elevated in patients with sporadic Alzheimer's disease and Down's syndrome. Ann Neurol 45: 537–541

Matsuzaki K, Horikiri C (1999) Interactions of amyloid β-peptide (1–40) with ganglioside-containing membranes. Biochemistry 38: 4137–4142

Mayeux R, Tang MX, Jacobs DM, Manly J, Bell K, Merchant C, Small SA, Stern Y, Wisniewski HM, Mehta PD (1999) Plasma amyloid β-peptide 1–42 and incipient Alzheimer's disease. Ann Neurol 46: 412–416

McGowan E, Sanders S, Iwatsubo T, Takeuchi A, Saido T, Zehr C, Yu X, Uljon S, Wang R, Mann D, Dickson D, Duff K (1999) Amyloid phenotype characterization of transgenic mice overexpressing both mutant amyloid precursor protein and mutant presenilin 1 transgenes. Neurobiol Dis 6: 231–244

McLean CA, Cherny RA, Fraser FW, Fuller SJ, Smith MJ, Beyreuther K, Bush AI, Masters CL (1999) Soluble pool of Aβ amyloid as a determinant of severity of neurodegeneration in Alzheimer's disease. Ann Neurol 46: 860–866

Mehta PD, Pirttila T, Mehta SP, Sersen EA, Aisen PS, Wisniewski HM (2000) Plasma and cerebrospinal fluid levels of amyloid β proteins 1–40 and 1–42 in Alzheimer's disease. Arch Neurol 57: 100–105

Moir RD, Lynch T, Bush AI, Whyte S, Henry A, Portbury S, Multhaup G, Small DH, Tanzi RE, Beyreuther K, Masters CL (1998) Relative increase in Alzheimer's disease of soluble forms of cerebral Aβ amyloid protein precursor containing the Kunitz protease inhibitory domain. J Biol Chem 273: 5013–5019

Moir RD, Atwood CS, Romano DM, Laurans MH, Huang X, Bush AI, Smith JD, Tanzi RE (1999) Differential effects of apolipoprotein E isoforms on metal-induced aggregation of Aβ using physiological concentrations. Biochemistry 38: 4595–4603

Mok SS, Sberna G, Heffernan D, Cappai R, Galatis D, Clarris HJ, Sawyer WH, Beyreuther K, Masters CL, Small DH (1997) Expression and analysis of heparin-binding regions of the amyloid precursor protein of Alzheimer's disease. FEBS Lett 415: 303–307

Morelli L, Giambartolomei GH, Prat MI, Castano EM (1999) Internalization and resistance to degradation of Alzheimer's Aβ1–42 at nanomolar concentrations in THP-1 human monocytic cell line. Neurosci Lett 262: 5–8

Multhaup G, Schlicksupp A, Hesse L, Beher D, Ruppert T, Masters CL, Beyreuther K (1996) The amyloid precursor protein of Alzheimer's disease in the reduction of copper (II) to copper. Science 271: 1406–1409

Multhaup G, Ruppert T, Schlicksupp A, Hesse L, Bill E, Pipkorn R, Masters CL, Beyreuther K (1998) Copper-binding amyloid precursor protein undergoes a site-specific fragmentation in the reduction of hydrogen peroxide. Biochemistry 37: 7224–7730

Murrell J, Farlow M, Ghetti B, Benson MD (1991) A mutation in the amyloid precursor protein associated with hereditary Alzheimer's disease. Science 254: 97–99

Näslund J, Haroutunian V, Mohs R, Davis KL, Greengard P, Buxbaum JD (2000) Correlation between elevated levels of amyloid β-peptide in the brain and cognitive decline. 283: 1571–1577

Nikaido T, Austin J, Rinehart R, Trueb L, Hutchinson J, Stukenbrok H, Miles B (1971) Studies in aging of the brain I. Isolation and preliminary characterization of Alzheimer plaques and cores. Archives of Neurology 25: 198–211

Niwa M, Sidrauski C, Kaufman RJ, Walter P (1999) A role for presenilin-1 in nuclear accumulation of Ire1 fragments and induction of the mammalian unfolded protein response. Cell 99: 691–702

Octave JN, Essalmani R, Tasiaux B, Menager J, Czech C, Mercken L (2000) The role of presenilin-1 in the γ-secretase cleavage of the amyloid precursor protein of Alzheimer's disease. J Biol Chem 275: 1525–1528

Palacino J, Berechid BE, Alexander P, Eckman C, Younkin S, Nye JS, Wolozin B (2000) Regulation of amyloid precursor protein processing by presenilin 1 (PS1) and PS2 in PS1 knockout cells. J Biol Chem 275: 215–222

Parkin ET, Turner AJ, Hooper NM (1999) Amyloid precursor protein, although partially detergent-insoluble in mouse cerebral cortex, behaves as an atypical lipid raft protein. Biochem J 344: 23–30

Peraus GC, Masters CL, Beyreuther K (1997) Late compartments of amyloid precursor protein transport in SY5Y cells are involved in β-amyloid secretion. J Neurosci 17: 7714–7724

Perez RG, Soriano S, Hayes JD, Ostaszewski B, Xia W, Selkoe DJ, Chen X, Stokin GB, Koo EH (1999) Mutagenesis identifies new signals for β-amyloid precursor protein endocytosis, turnover, and the generation of secreted fragments, including Aβ42. J Biol Chem 274: 18851–18856

Pillot T, Drouet B, Queille S, Labeur C, Vandekerckhove J, Rosseneu M, Pincon-Raymond M, Chambaz J (1999a) The nonfibrillar amyloid β-peptide induces apoptotic neuronal cell death: involvement of its C-terminal fusogenic domain. J Neurochem 73: 1626–1634

Pillot T, Goethals M, Najib J, Labeur C, Lins L, Chambaz J, Brasseur R, Vandekerckhove J, Rosseneu M (1999b) β-amyloid peptide interacts specifically with the carboxy-terminal domain of human apolipoprotein E: relevance to Alzheimer's disease. J Neurochem 72: 230–237

Poduslo JF, Curran GL, Sanyal B, Selkoe DJ (1999) Receptor-mediated transport of human amyloid β-protein 1–40 and 1–42 at the blood-brain barrier. Neurobiol Dis 6: 190–199

Ponte P, Gonzalez-DeWhitt P, Schilling J, Miller J, Hsu D, Greenberg B, Davis K, Wallace W, Lieberburg I, Fuller F (1988) A new A4 amyloid mRNA contains a domain homologous to serine proteinase inhibitors. Nature 331: 525–527

Pradier L, Carpentier N, Delalonde L, Clavel N, Bock MD, Buee L, Mercken L, Tocque B, Czech C (1999) Mapping the APP/presenilin (PS) binding domains: the hydrophilic N-terminus of PS2 is sufficient for interaction with APP and can displace APP/PS1 interaction. Neurobiol Dis 6: 43–55

Qiu Z, Strickland DK, Hyman BT, Rebeck GW (1999) A2-macroglobulin enhances the clearance of endogenous soluble β-amyloid peptide via low-density lipoprotein receptor-related protein in cortical neurons. J Neurochem 73: 1393–1398

Quon D, Wang Y, Catalano R, Scardina JM, Murakami K, Cordell B (1991) Formation of β-amyloid protein deposits in brains of transgenic mice. Nature 352: 239–241

Ray WJ, Yao M, Mumm J, Schroeter EH, Saftig P, Wolfe M, Selkoe DJ, Kopan R, Goate AM (1999) Cell surface presenilin-1 participates in the γ-secretase-like proteolysis of Notch. J Biol Chem 274: 36801–36807

Redlich E (1898) Ueber miliäre Skerose der Hirnrinde bei seniler Atrophie. Jahrb Psychiat Neurol 17: 208

Robakis NK, Ramakrishna N, Wolfe G, Wisniewski HM (1987) Molecular cloning and characterization of a cDNA encoding the cerebrovascular and the neuritic plaque amyloid peptides. Proc Natl Acad Sci USA 84: 4190–4194

Rogaev EI, Sherrington R, Rogaeva EA, Levesque G, Ikeda M, Liang Y, Chi H, Lin C, Holman K, Tsuda T, Mar L, Sorbi S, Nacmias B, Piacentini S, Amaducci L, Chumakov I, Cohen D, Lannfelt L, Fraser PE, Rommens JM, St George-Hyslop PH (1995) Familial Alzheimer's disease in kindreds with missense mutations in a gene on chromosome 1 related to the Alzheimer's disease type 3 gene. Nature 376: 775–778

Roher A, Wolfe D, Palutke M, KuKuruga D (1986) Purification, ultrastructure, and chemical analysis of Alzheimer disease amyloid plaque core protein. Proc Natl Acad Sci USA 83: 2662–2666

Roher AE, Kuo YM, Kokjohn KM, Emmerling MR, Gracon S (1999) Amyloid and lipids in the pathology of Alzheimer's disease. Amyloid: Int J Exp Clin Invest 6: 136–145

Samuels SC, Silverman JM, Marin DB, Peskind ER, Younki SG, Greenberg DA, Schnur E, Santoro J, Davis KL (1999) CSF β-amyloid, cognition, and APOE genotype in Alzheimer's disease. Neurology 52: 547–551

Sandbrink R, Masters CL, Beyreuther K (1994a) Complete nucleotide and deduced amino acid sequence of rat amyloid protein precursor-like protein (APLP2/APPH): two amino acids length difference to human and murine homologues. Biochim Biophys Acta 1219: 167–170

Sandbrink R, Masters CL, Beyreuther K (1994b) Similar alternative splicing of a non-homologous domain in βA4-amyloid protein precursor-like proteins. J Biol Chem 269: 14227–14234

Sberna G, Saez-Valero J, Beyreuther K, Masters CL, Small DH (1997) The amyloid β-protein of Alzheimer's disease increases acetylcholinesterase expression by increasing intracellular calcium in embryonal carcinoma P19 cells. J Neurochem 69: 1177–1184

Sberna G, Saez-Valero J, Li QX, Czech C, Beyreuther K, Masters CL, McLean CA, Small DH (1998) Acetylcholinesterase is increased in the brains of transgenic mice expressing the C-terminal fragment (CT100) of the β-amyloid protein precursor of Alzheimer's disease. J Neurochem 71: 723–731

Scharnagl H, Tisljar U, Winkler K, Huttinger M, Nauck MA, Gross W, Wieland H, Ohm TG, Marz W (1999) The βA4 amyloid peptide complexes to and enhances the uptake of β-very low density lipoproteins, the low density lipoprotein receptor-related protein and heparan sulfate proteoglycans pathway. Lab Invest 79: 1271–1286

Schenk D, Barbour R, Dunn W, Gordon G, Grajeda H, Guido T, Hu K, Huang J, Johnson-Wood K, Khan K, Kholodenko D, Lee M, Liao Z, Lieberburg I, Motter R, Mutter L, Soriano F, Shopp G, Vasquez N, Vandevert C, Walker S, Wogulis M, Yednock T, Games D, Seubert P (1999) Immunization with amyloid-β attenuates Alzheimer-disease-like pathology in the PDAPP mouse. Nature 400: 173–177

Scheuner D, Eckman C, Jensen M, Song X, Citron M, Suzuki N, Bird TD, Hardy J, Hutton M, Kukull W, Larson E, Levy-Lahad E, Viitanen M, Peskind E, Poorkaj P, Schellenberg G, Tanzi R, Wasco W, Lannfelt L, Selkoe D, Younkin S (1996) Secreted amyloid β-protein similar to that in the senile plaques of Alzheimer's disease is increased in vivo by the presenilin 1 and 2 and APP mutations linked to familial Alzheimer's disease. Nat Med 2: 864–870

Scholz W (1938) Studien zur Pathologie der Hirngefäße II. Die drusige Entartung der Hirnarterien und -capillaren. (Eine Form seniler Gefäßerkrankung.) Zeit ges Neurol Psychiat 162: 694–715

Schröder J, Pantel J, Ida N, Essig M, Hartmann T, Knopp MV, Schad LR, Sandbrink R, Sauer H, Masters CL, Beyreuther K (1997) Cerebral changes and cerebrospinal fluid β-amyloid in Alzheimer's disease – a study with quantitative magnetic resonance imaging. Mol Psychiat 2: 505–507

Schubert D, Chevion M (1995) The role of iron in β amyloid toxicity. Biochem Biophys Res Commun 216: 702–707

Schwarzman AL, Singh N, Tsiper M, Gregori L, Dranovsky A, Vitek MP, Glabe CG, St George-Hyslop PH, Goldgaber D (1999) Endogenous presenilin 1 redistributes to the surface of lamellipodia upon adhesion of Jurkat cells to a collagen matrix. Proc Natl Acad Sci USA 96: 7932–7937

Sekijima Y, Kametani F, Tanaka K, Okochi M, Usami M, Mori H, Tokuda T, Ikeda S (1999) Presenilin-1 exists in the axoplasm fraction in the brains of aged Down's syndrome subjects and non-demented individuals. Neurosci Lett 267:121–124

Selkoe DJ (1999) Translating cell biology into therapeutic advances in Alzheimer's disease. Nature 399: A23–31

Selkoe DJ, Abraham CR, Podlisny MB, Duffy LK (1986) Isolation of low-molecular-weight proteins from amyloid plaque fibers in Alzheimer's disease. J Neurochem 46: 1820–1834

Seubert P, Vigo-Pelfrey C, Esch F, Lee M, Dovey H, Davis D, Sinha S, Schlossmacher M, Whaley J, Swindleshurst C, McCormack R, Wolfert R, Selkoe D, Lieberburg I, Schenk D (1992) Isolation and quantification of soluble Alzheimer's β-peptide from biological fluids. Nature 359: 325–327

Seubert P, Oltersdorf T, Lee MG, Barbour R, Blomquist C, Davis DL, Bryant K, Fritz LC, Galasko D, Thal LJ, Lieberburg I, Schenk DB (1993) Secretion of β-amyloid precursor protein cleaved at the amino terminus of the β-amyloid peptide. Nature 361: 260–263

Shao H, Jao S, Ma K, Zagorski MG (1999) Solution structures of micelle-bound amyloid β-(1–40) and β-(1–42) peptides of Alzheimer's disease. J Mol Biol 285: 755–773

Sherrington R, Rogaev EI, Liang Y, Rogaeva EA, Levesque G, Ikeda M, Chi H, Lin C, Li G, Holman K, Tsuda T, Mar L, Foncin JF, Bruni AC, Montesi MP, Sorbi S, Rainero I, Pinnessi L, Nee L, Chumakov I, Pollen D, Brookes A, Sanseau P, Polinsky RJ, Wasco W, Da Silva HAR, Haines JL, Pericak-Vance MA, Tanzi RE, Roses AD, Rommens JM, St George Hyslop PH (1995) Cloning of a gene bearing missense mutations in early-onset familial Alzheimer's disease. Nature 375: 754–760

Shirotani K, Takahashi K, Araki W, Maruyama K, Tabira T (2000) Mutational analysis of intrinsic regions of presenilin 2 that determine its endoproteolytic cleavage and pathological function. J Biol Chem 275: 3681–3686

Shoji M, Golde TE, Ghiso J, Cheung TT, Estus S, Shaffer LM, Cai XD, McKay DM, Tintner R, Frangione B, Younkin SG (1992) Production of the Alzheimer amyloid β protein by normal proteolytic processing. Science 258: 126–129

Sinha S, Anderson JP, Barbour R, Basi GS, Caccavello R, Davis D, Doan M, Dovey HF, Frigon N, Hong J, Jacobson-Croak K, Jewett N, Keim P, Knops J, Lieberburg I, Power M, Tan H, Tatsuno G, Tung J, Schenk D, Seubert P, Suomensaari SM, Wang S, Walker D, Zhao J, McConlogue L, John V (1999) Purification and cloning of amyloid precursor protein β-secretase from human brain. Nature 402: 537–540

Skovronsky DM, Moore DB, Milla ME, Doms RW, Lee VM (2000) Protein kinase C-dependent α-secretase competes with β-secretase for cleavage of amyloid-β precursor protein in the trans-Golgi network. J Biol Chem 275: 2568–2575

Soriano S, Chyung AS, Chen X, Stokin GB, Lee VM, Koo EH (1999) Expression of β-amyloid precursor protein-CD3γ chimeras to demonstrate the selective generation of amyloid β (1–40) and amyloid β (1–42) peptides within secretory and endocytic compartments. J Biol Chem 274: 32295–32300

Steiner H, Romig H, Pesold B, Philipp U, Baader M, Citron M, Loetscher H, Jacobsen H, Haass C (1999) Amyloidogenic function of the Alzheimer's disease-associated presenilin 1 in the absence of endoproteolysis. Biochemistry 38: 14600–14605

Storey E, Spurck T, Pickett-Heaps J, Beyreuther K, Masters CL (1996a) The amyloid precursor protein of Alzheimer's disease is found on the surface of static but not activity motile portions of neurites. Brain Res 735: 59–66

Storey E, Beyreuther K, Masters CL (1996b) Alzheimer's disease amyloid precursor protein on the surface of cortical neurons in primary culture co-localizes with adhesion patch components. Brain Res 735: 217–231

Storey E, Katz M, Brickman Y, Beyreuther K, Masters CL (1999) Amyloid precursor protein of Alzheimer's disease: evidence for a stable, full-length, trans-membrane pool in primary neuronal cultures. Eur J Neurosci 11: 1779–1788

Strittmatter WJ, Weisgraber KH, Huang DY, Dong LM, Salvesen GS, Pericak-Vance M, Schmechel D, Saunders AM, Goldgaber D, Roses AD (1993) Binding of human apolipoprotein E to synthetic amyloid β peptide: isoform-specific effects and implications for late-onset Alzheimer disease. Proc Natl Acad Sci USA 90: 8098–8102

Sturchler-Pierrat C, Abramowski D, Duke M, Wiederhold KH, Mistl C, Rothacher S, Ledermann B, Burki K, Frey P, Paganetti PA, Waridel C, Calhoun ME, Jucker M, Probst A, Staufenbiel M, Sommer B (1997) Two amyloid precursor protein transgenic mouse models with Alzheimer disease-like pathology. Proc Natl Acad Sci USA 94: 13287–13292

Tanzi RE, Gusella JF, Watkins PC, Bruns GAP, St George-Hyslop P, van Keuren ML, Patterson D, Pagan S, Kurnit DM, Neve RL (1987) Amyloid β protein gene: cDNA, mRNA distribution and genetic linkage near the Alzheimer locus. Science 235: 880–884

Tanzi RE, McClatchey AI, Lamperti ED, Villa-Komaroff L, Gusella JF, Neve RL (1988) Protease inhibitor domain encoded by an amyloid protein precursor mRNA associated with Alzheimer's disease. Nature 331: 528–530

Tapiola T, Pirttila T, Mikkonen M, Mehta PD, Alafuzoff I, Koivisto K, Soininen H (2000) Three-year follow-up of cerebrospinal fluid tau, β-amyloid 42 and 40 concentrations in Alzheimer's disease. Neurosci Lett 280: 119–122

Tienari PJ, De Strooper B, Ikonen E, Simons M, Weidemann A, Czech C, Hartmann T, Ida N, Multhaup G, Masters CL, Van Leuven F, Beyreuther K, Dotti CG (1996a) The β-amyloid domain is essential for axonal sorting of amyloid precursor protein. EMBO J 15: 5218–5229

Tienari PJ, De Strooper B, Ikonen E, Ida N, Simons M, Masters CL, Dotti CG, Beyreuther K (1996b) Neuronal sorting and processing of amyloid precursor protein: implications for Alzheimer's disease. Cold Spring Harb Symp Quant Biol 61: 575–585

Tienari PJ, Ida N, Ikonen E, Simons M, Weidemann A, Multhaup G, Masters CL, Dotti CG, Beyreuther K (1997) Intracellular and secreted Alzheimer's β-amyloid species are generated by distinct mechanisms in cultured hippocampal neurons. Proc Natl Acad Sci USA 94: 4125–4130

Tjernberg LO, Callaway DJ, Tjernberg A, Hahne S, Lilliehook C, Terenius L, Thyberg J, Nordstedt C (1999) A molecular model of Alzheimer's amyloid β-peptide fibril formation. J Biol Chem 274: 12619–12625

Vassar R, Bennett BD, Babu-Khan S, Kahn S, Mendiaz EA, Denis P, Teplow DB, Ross S, Amarante P, Loeloff R, Luo Y, Fisher S, Fuller J, Edenson S, Lile J, Jarosinski MA, Biere AL, Curran E, Burgess T, Louis JC, Collins F, Treanor J, Rogers G, Citron M (1999) β-secretase cleavage of Alzheimer's amyloid precursor protein by the transmembrane aspartic protease BACE. Science 286: 735–741

Vekrellis K, Ye Z, Qiu WQ, Walsh D, Hartley D, Chesneau V, Rosner MR, Selkoe DJ (2000) Neurons regulate extracellular levels of amyloid β-protein via proteolysis by insulin-degrading enzyme. J Neurosci 20: 1657–1665

Walsh DM, Hartley DM, Kusumoto Y, Fezoui Y, Condron MM, Lomakin A, Benedek GB, Selkoe DJ, Teplow DB (1999) Amyloid β-protein fibrillogenesis. Structure and biological activity of protofibrillar intermediates. J Biol Chem 274: 25945–25952

Wang J, Dickson DW, Trojanowski JQ, Lee VM (1999) The levels of soluble versus insoluble brain Aβ distinguish Alzheimer's disease from normal and pathologic aging. Exp Neurol 158: 328–337

Weidemann A, Paliga K, Durrwang U, Czech C, Evin G, Masters CL, Beyreuther K (1997) Formation of stable complexes between two Alzheimer's disease gene products: presenilin-2 and β-amyloid precursor protein. Nature Med 3: 328–332

Weidemann A, Paliga K, Dürrwang U, Reinhard FBM, Schuckert O, Evin G, Masters CL (1999) Proteolytic processing of the Alzheimer's disease amyloid precursor protein within its cytoplasmic domain by caspase-like proteases. J Biol Chem 274: 5823–5829

Whitson JS, Selkoe DJ, Cotman CW (1989) Amyloid β protein enhances the survival of hippocampal neurons in vitro. Science 243: 1488–1490

Whyte S, Wilson N, Currie J, Maruff P, Malone V, Shafiq-Antonacci R, Tyler P, Derry KL, Underwood J, Li QX, Beyreuther K, Masters CL (1997) Collection and normal levels of the amyloid precursor protein in plasma. Ann Neurol 41: 121–124

Wickner S, Maurizi MR, Gottesman S (1999) Posttranslational quality control: folding, refolding, and degrading proteins. Science 286: 1888–1893

Wisniewski T, Frangione B (1992) Apolipoprotein E: a pathological chaperone protein in patients with cerebral and systemic amyloid. Neurosci Lett 135: 235–238

Williamson TG, Mok SS, Henry A, Cappai R, Lander AD, Nurcombe V, Beyreuther K, Masters CL, Small DH (1996) Secreted glypican binds to the amyloid precursor protein of Alzheimer's disease (APP) and inhibits APP-induced neurite outgrowth. J Biol Chem 271: 31215–31221

Wilson CA, Doms RW, Lee VM (1999) Intracellular APP processing and Aβ production in Alzheimer disease. J Neuropathol Exp Neurol 58: 787–794

Wolfe MS, Xia W, Ostaszewski BL, Diehl TS, Kimberly WT, Selkoe DJ (1999a) Two transmembrane aspartates in presenilin-1 required for presenilin endoproteolysis and γ-secretase activity. Nature 398: 513–517

Wolfe MS, Xia W, Moore CL, Leatherwood DD, Ostaszewski B, Rahmati T, Donkor IO, Selkoe DJ (1999b) Peptidomimetic probes and molecular modeling suggest that Alzheimer's γ-secretase is an intramembrane-cleaving aspartyl protease. Biochemistry 38: 4720–4727

Yamin R, Malgeri EG, Sloane JA, McGraw WT, Abraham CR (1999) Metalloendopeptidase EC 3.4.24.15 is necessary for Alzheimer's amyloid-β peptide degradation. J Biol Chem 274: 18777–18784

Yan R, Bienkowski MJ, Shuck ME, Miao H, Tory MC, Pauley AM, Brashier JR, Stratman NC, Mathews WR, Buhl AE, Carter DB, Tomasselli AG, Parodi LA, Heinrikson RL, Gurney ME (1999) Membrane-anchored aspartyl protease with Alzheimer's disease β-secretase activity. Nature 402: 533–537

Yang AJ, Chandswangbhuvana D, Shu T, Henschen A, Glabe CG (1999a) Intracellular accumulation of insoluble, newly synthesized Aβn-42 in amyloid precursor protein-transfected cells that have been treated with Aβ1–42. J Biol Chem 274: 20650–20656

Yang DS, Small DH, Seydel U, Smith JD, Hallmayer J, Gandy SE, Martins RN (1999b) Apolipoprotein E promotes the binding and uptake of β-amyloid into Chinese hamster ovary cells in an isoform-specific manner. Neuroscience 90: 1217–1226

Yang DS, Yip CM, Huang TH, Chakrabartty A, Fraser PE (1999c) Manipulating the amyloid-β aggregation pathway with chemical chaperones. J Biol Chem 274: 32970–32974

Yankner BA, Dawes LR, Fisher S, Villa-Komaroff L, Oster-Granite ML, Neve RL (1989) Neurotoxicity of a fragment of the amyloid precursor associated with Alzheimer's disease. Science 245: 417–420

Mechanisms of Motor Neuron Death in ALS

T. L. Williamson and D. W. Cleveland

Introduction

Amyotrophic lateral sclerosis (ALS) is a fatal disease in which degeneration of upper and lower motor neurons leads to progressive muscle weakness. The most common form of adult motor neuron disease in humans, ALS typically initiates in middle to late life, leading to paralysis and death within three to five years. The disease usually begins asymmetrically in one limb, most commonly the leg, and then appears to spread to involve contiguous groups of motor neurons. Approximately 10 % of ALS cases are inherited in an autosomal dominant fashion. Many affected neurons in ALS patients show cytoskeletal pathology in the form of neurofilament accumulations, both within the cell bodies and in proximal axons (Carpenter 1968; Chou and Fakadej 1971; Hirano et al. 1984a, b). Motor neuron loss is also accompanied by reactive gliosis (Leigh et al. 1991), ubiquitin-positive inclusions (Leigh et al. 1991) and apparent fragmentation of the Golgi (Gonatas et al. 1992). A landmark discovery in deciphering the mechanism of the disease came from the identification of the mutations in the gene encoding cytoplasmic superoxide dismutase (SOD1) that underlie about 20 % of the instances of inherited disease (Rosen et al. 1993; Siddique and Deng 1996; Andersen et al. 2000).

Four primary hypotheses for contributors to the disease mechanism have now emerged: 1) oxidative damage, an idea obviously provoked by the discovery of SOD1 mutations; 2) axonal strangulation from neurofilamentous disorganization, supported by misaccumulated neurofilaments as a hallmark of pathology in many cases of sporadic and SOD1 mediated familial disease; 3) toxicity arising from intracellular aggregates and/or failure of protein folding or degradation, a common feature of SOD1 mutant-mediated disease; and 4) excitotoxicity from aberrant handling of glutamate, particularly arising from missplicing of a glutamate transporter mRNA. We provide here a status report on the contribution of the first three of these. A discussion of errors of glutamate metabolism can be found elsewhere (Cleveland 1999).

SOD1 and ALS

The only proven primary cause of ALS is mutation in SOD1, a copper- and zinc-containing enzyme of 153 amino acids that catalyzes the production of oxygen or hydrogen peroxide from superoxide anions through two, asymmetric catalytic

Research and Perspectives in Alzheimer's Diseases
Beyreuther/Christen/Masters (Eds.)
Neurodegenerative Disorders
© Springer-Verlag Berlin Heidelberg 2001

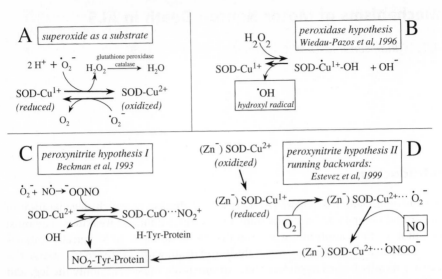

Fig. 1. Normal and proposed aberrant SOD1 chemistries. (A) Normal SOD1 chemistry: SOD1-mediated dismutation of superoxide in two asymmetric steps. (B–D) Proposed toxic chemistries from use of aberrant substrates. (B) Hydroxyl radical formation from use of hydrogen peroxide as a substrate by the Cu[1+] oxidation form of SOD1. (C,D) Proposed protein nitration from either of two different pathways for use of peroxynitrite (ONOO) as an aberrant substrate. (C) Accelerated use of spontaneously formed peroxynitrite by the mutant SOD1. (D) Reduced zinc binding by mutant SOD1 allows rapid intracellular reduction to the Cu[1+] form. This enzyme form then runs catalysis backward to produce superoxide, which in turn combines within the mutant active site with freely diffusing nitric oxide (NO) to produce peroxynitrite

steps (Fig. 1A) in which the active site copper is successively oxidized, then reduced, thereby switching from Cu^{1+} to Cu^{2+} and back (Fridovich 1995). SOD1 is thought to be expressed in all eukaryotic cells and its activity is believed to play a crucial role in protection against oxygen radical-induced cellular damage (Halliwell 1994; Yu 1994). It is especially abundant in nervous tissues, comprising between 0.1 and 1 % of total brain or spinal cord proteins, making it as abundant in those tissues as some of the major cytoskeletal proteins.

Transgenic Mice Models of SOD1-Mediated Motor Neuron Disease

With the discovery of mutations in SOD1 as a primary cause of disease, the key question is how do mutations in SOD1 (a highly abundant, ubiquitous enzyme) lead to late onset, selective motor neuron degeneration? Despite an early focus on the proposal that dominantly inherited disease may arise from loss of enzymatic activity (McNamara and Fridovich 1993; Deng et al. 1993; Robberecht et al. 1994; Orrell at al. 1995), no correlation exists between retention of enzymatic activity by mutants (Borchelt et al. 1994, 1995) and the timing or disease onset or rapidity of disease progression (Cleveland et al. 1995). Most compelling, the production of SOD1 mutant-expressing transgenic mouse models, along with analysis of mice in which the SOD1 gene was deleted, have combined to demonstrate that the mutants confer disease from an acquired toxic property unlinked to SOD1 activity.

In the first of these models, Gurney et al. (1994) produced transgenic mice expressing the familial ALS-linked mutation SOD1^{G93A} (that is glycine substituted to alanine at position 93) which developed motor neuron degeneration that both clinically and pathologically resembled ALS. The affected SOD1^{G93A} mice became paralyzed in one or more limbs as a result of the loss of spinal cord motor neurons and died at ~5–6 months of age. Overexpression of wild type human SOD1 generated no phenotype. The pathological changes in the SOD1^{G93A} line have now been documented extensively. Abnormally phosphorylated neurofilaments (Gurney et al. 1994) and accompanying neurofilament aggregates (Tu et al. 1996) were initially reported in the cell bodies of most of the remaining motor neurons at end stage, but these are less striking than reported in human disease. The earliest and most prominent change of large motor neurons in the spinal cord in the SOD1^{G93A} mice is microvesiculation, apparently originating from dilation of rough endoplasmic reticulum and degenerating mitochondria (Dal Canto and Gurney 1994, 1995, 1997).

Rapidly following the report of disease in an initial SOD1^{G93A} mutant mouse line was an analysis of a set of transgenic mice accumulating another of the familial ALS-associated mutant protein (SOD1^{G37R}). As this mutant retains full specific activity and was shown to be present with only a modestly reduced level in lymphocytes from patients (Borchelt et al. 1994), it seemed at the outset highly unlikely that this mutant disease could arise from loss of activity. Indeed, four lines of mice expressing the mutant between 4 and 12 times the level of the endogenous mouse SOD1 polypeptide in spinal cord each developed fatal, progressive motor neuron disease (Wong et al. 1995). At lower levels of mutant accumulation, pathology was restricted to lower motor neurons, whereas higher levels caused more severe abnormalities and affected a variety of other neuronal populations. Like the case for mice expressing SOD1^{G93A}, the most obvious cellular abnormality was the presence in axons and dendrites of membrane-bounded vacuoles, which appeared at the electron microscopic level to be derived from degenerating mitochondria. However, in the SOD1^{G37R} mice the vacuoles were strongly enriched in the proximal axons and dendrites with the neuronal cell bodies were largely spared, in contrast with SOD1^{G93A} mice in which the motor neuron cell bodies were primarily affected (Dal Canto and Gurney 1994). Since multiple lines of mice expressing wild-type human SOD1 at up to 14 times normal mouse SOD1 polypeptide levels (with a corresponding increase in SOD1 activity) do not develop an overt phenotype and complete deletion of SOD1 yields neither overt disease nor compromised life span (Reaume et al. 1996), the disease in mice expressing the SOD1^{G37R} mutant SOD1 must arise from the acquisition of an adverse property by the mutant enzyme, rather than elevation or loss of SOD1 activity.

In both the SOD1^{G93A} and SOD1^{G37R} mouse models of familial ALS-linked SOD1 mutations, very high levels (more than four times the endogenous SOD1) of mutant proteins were required to provoke disease. This finding raised an obvious question about the degree to which a disease requiring these excessive levels of mutant subunits was faithfully mimicking the human disease pathway. An initial reassurance arose from the report (Ripps et al. 1995) of one mouse line that developed a very rapidly progressing motor neuron disease from expressing a mutation (SOD1^{G86R}) in mouse SOD1 that is homologous in position to the familial ALS-linked human SOD1^{G85R} mutation. Disease arose in this mouse without apparent elevation or loss of SOD1 activity. This finding was significantly extended by analysis of a set of 25 founder mice accumulating relatively low levels of human SOD1 mutant SOD1^{G85R} (Bruijn et al. 1997b). Seven lines of mice that accumulated this mutant at levels between 20% and 100% of the endogenous wild type mouse SOD1 (e.g., levels appropriate for dominantly inherited disease) invariably developed motor neuron disease characterized by an extremely rapid clinical progression after onset.

The most striking pathologic finding in transgenic mice expressing human SOD1^{G85R} is the presence of numerous inclusions, which immunostaining with antibodies to GFAP revealed to be within astrocytes. These inclusions appear prior to, or contemporaneous with, similar aggregates in motor neurons. By end-stage, when there is a two-fold elevation in SOD1 protein in whole spinal cord extracts and a 60% loss of large axons, inclusions are approximately 10 times more abundant in astrocytes than in neurons. Strikingly, both the dense core and the periphery of both astrocytic and neuronal inclusions are intensely reactive with antibodies to SOD1. Similar neuronal SOD1-positive inclusions have been described in some familial ALS patients with mutation SOD1^{A4V} (Shibata et al. 1996), and inclusions with similar SOD1 immunoreactivity in astrocytes have also been reported (Kato et al. 1997) in a familial ALS patient with a two base pair deletion in the 126th codon of SOD1, leading to a frameshift and truncation of the final 27 amino acids (Kato et al. 1996). The demonstration that SOD1-containing inclusions in astrocytes are early abnormalities and increase markedly in abundance after disease onset provides support for the view that at least some of the familial ALS-linked SOD1 mutants may mediate disease by direct effects on astrocytes, in both human and murine disease.

Complete Deletion of the SOD1 Gene Does not Cause Motor Neuron Disease in Mice

Further evidence showing that ALS mediated by SOD1 mutation arises from the gain of a toxic property(ies) of the mutants was provided through the use of homologous recombination to produce a complete deletion of the SOD1 gene in mice (Reaume et al. 1996). While SOD1 is essential for efficient aerobic growth of yeast (Liu et al. 1992), mice completely deficient in SOD1 develop normally without overt phenotype. These mice do, however, show reduced motor unit numbers early in life, followed by a slow decline with age, suggesting that some axonal sprouting and reinnervation of denervated muscle fibers continues during aging in the chronic, complete absence of SOD1 (Shefner et al. 1999). SOD1-deficient mice did exhibit marked vulnerability to motor neuron loss after axonal injury. Nevertheless, these mice showed no changes in stride length or running activity and showed no tremor or paralysis typical of the SOD1 mutant mice. No vacuolization, cell loss or other pathological changes were observed in the brain or spinal cord. Brain tissues from SOD1-deficient mice did not show any increases in either lipid peroxidation or protein carbonyl content, two common end products of oxidative damage.

The absence of overt disease in mice completely lacking SOD1, combined with the progressive disease arising in mice expressing familial ALS-linked mutations SOD1^{G93A} and SOD1^{G37R} (which have significantly elevated SOD1 activity levels) or SOD1^{G85R} (which have normal SOD1 activity levels), demonstrates that disease in mice arises from a toxic property of the mutant subunits rather than a loss of SOD1 activity. Moreover, matings between SOD1^{G85R} mice with either SOD1 null mice or mice overexpressing high levels of human SOD1 had no effect on disease onset or progression (Bruijn et al. 1998), findings that call into question whether toxicity is related to metabolism of superoxide or any spontaneous reaction product of it.

Identifying the Toxic Property of the Familial ALS-SOD1 Mutants

The obvious question emerging from these initial transgenic mouse efforts is, what is the toxic property(ies) of the familial ALS-linked SOD1 mutants? One initial proposed mechanism (Fig. 1C) is that the mutant subunits unfold slightly, allowing increased access of peroxynitrite (the spontaneous reaction product of superoxide and nitric oxide) to the enzyme-bound copper, which in turn uses peroxynitrite to nitrate tyrosine residues on proteins (Beckman et al. 1993). A competing hypothesis (Wiedau-Pazos et al. 1996) is that mutants catalyze the formation of hydroxyl radicals from hydrogen peroxide via the Fenton reaction (Fig. 1B). With subsequent evidence arguing strongly against both of these hypotheses (documented below), a modified peroxynitrite proposal has emerged from Beckman and colleagues (Estevez et al. 1999) in which the mutant SOD1 produce peroxynitrite within their active sites (Fig. 1D). As we summarize below, this hypothesis, too, seems unlikely to be correct, since it is at odds with most in vivo evidence.

The Initial Peroxynitrite Hypothesis

Initial evidence in favor of the first peroxynitrite proposal was provided by immunostaining with an antibody to nitrotyrosine: this was reported to yield increased staining in motor neurons from sporadic ALS spinal cords (Abe et al. 1995, 1997; Chou et al. 1996; Beal et al. 1997) and in the SOD1^{G93A} transgenic mouse model (Ferrante et al. 1997). However, subsequent efforts with other transgenic mouse models demonstrated that this hypothesis cannot be a general property of the ALS-linked SOD1 mutants. A similar immunocytochemical examination for nitrotyrosine in two lines of transgenic mice that develop progressive motor neuron disease from expressing human familial ALS-linked SOD1 mutation SOD1^{G37R} (Bruijn et al. 1997a) and in one line expressing SOD1^{G85R} (Bruijn et al. 1998) found no detectable protein-bound nitrotyrosine at any stage of disease progression. The inability to detect nitration even of highly abundant proteins such as neurofilaments and GFAP, despite a sensitivity capable of detecting signals from as little as 2 ng of in vitro nitrated standards, indicates that the stoichiometry of protein-bound nitration must be very low, if present at all.

More conclusive evidence that this first peroxynitrite hypothesis could not be correct arose from testing how the presence or absence of wild type SOD1 affected disease produced by the nearly inactive mutant SOD1^{G85R} [which retains only about 10 % of the activity of wild type SOD1 (Corson et al. 1998) and causes disease even at only 20 % of the level of the endogenous SOD1 (Bruijn et al. 1997a)]. If this mutant used peroxynitrite as an aberrant substrate more efficiently than did wild type SOD1, then disease would be exacerbated by removing the endogenous, wild type SOD1, since this removal would send all superoxide and/or peroxynitrite catalysis through the mutant, thereby accelerating damage. Moreover, a chronic increase in wild type SOD1 (by mating the SOD1^{G85R} transgenic line to one carrying a wild type SOD1 transgene) would ameliorate disease by competition of the more abundant (and more active) wild type enzyme for the available substrates. However, the outcome of this pair of experiments (Fig. 2) demonstrated unambiguously that neither elimination nor elevation of the wild type SOD1 affected the timing or course of the disease (Bruijn et al. 1998), an outcome inconsistent with this initial peroxynitrite hypothesis.

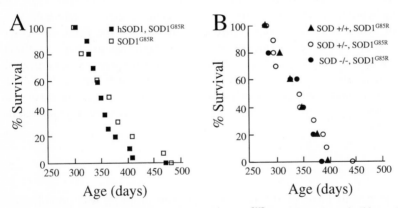

Fig. 2. Disease onset and progression mediated by SOD1^{G85R} are independent of wild type SOD1 protein or activity levels. (A) Survival plots of SOD1^{G85R} transgenic mice with (■) or without (□) a human wild type SOD1 transgene to elevate wild type SOD1 levels to six times the normal amount. (B) Survival plots of mice with the SOD1^{G85R} transgene and both (▲) endogenous mouse SOD1 genes or heterozygously (○) or homozygously (●) deleted for endogenous mouse SOD1. Reprinted with permission from Bruijn et al. (1998)

The Peroxide Hypothesis

To address the question of whether mutant SOD1 acts as a peroxidase, initial measurements for predicted oxidative products were undertaken in samples from both sporadic and familial ALS. Oddly, despite apparent increases in oxidative markers in sporadic disease, no elevations were identified in familial disease, including that arising from mutation in SOD1 (Bowling et al. 1993). Salicylate trapping was then used to measure the levels of hydroxyl radicals in mice expressing the SOD1^{G37R} mutation (Bruijn et al. 1997a). This revealed no difference in hydroxyl radical levels in comparing SOD1^{G37R} mice with control animals. Surprisingly, a significant increase in salicylate reaction products was reproducibly found in mice expressing high levels of wild type human SOD1. Spin trapping confirmed the elevated levels of hydroxyl radicals in a different line of mice expressing excessive wild type SOD1 (Peled-Kamar et al. 1997). While this result is certainly unexpected, the presence of ALS-like disease in mice expressing mutant SOD1 in the absence of elevated hydroxyl radicals, strongly suggests that hydroxyl radicals are not the toxic products of familial ALS-linked SOD1 mutants. In addition, a search for lipid peroxidation by measuring levels of the peroxidation product malondialdehyde (MDA; Bruijn et al. 1997a) revealed no increase in MDA levels in SOD1^{G37R} mice, thus offering no support for increased lipid peroxidation resulting from the mutant human SOD1. Along with a direct challenge on technical grounds of the initial evidence interpreted in support of the peroxide hypothesis (Singh et al. 1998), there remains little likelihood that this hypothesis can explain a toxicity common to the ALS-linked SOD1 mutants.

A Second Peroxynitrite Hypothesis: Zinc-Deficient SOD1 Runs Catalysis Backwards to Generate Peroxynitrate Within its Catalytic Site

A more recent report, again from the Beckman group (Estevez et al. 1999), yielded a second peroxynitrite hypothesis (Fig. 1D), in which relative to the wild type SOD1, the mutant subunits were proposed to fail to bind or retain the zinc atom, thereby allowing rapid reduction of mutant SOD1 to the Cu^{1+} form by abundant intracellular reductants. The reduced SOD1 mutant would then run the normal catalytic step backwards, converting oxygen to superoxide. Finally, the superoxide so produced would then combine within the enzymatic active site with freely diffusing nitric oxide (NO), thereby producing peroxynitrite, which would promote intracellular damage including protein nitration (Estevez et al. 1999). The primary evidence in support of this view was that introduction by liposome fusion of purified SOD1 that had been depleted of zinc but was still loaded with copper provoked rapid death of cultured motor neurons. Toxicity required both zinc depletion and bound copper.

The evidence was persuasive concerning cell death in vitro but, as initially pointed out by Williamson et al. (2000), this acute toxicity of zinc-free wild type or mutant SOD1 seems likely to have little in common with the in vivo pathway of motor neuron death arising from chronic expression of the mutants over a slow time course.

First, as noted by Estevez et al. (1999), zinc-depleted wild type were just as toxic as the ALS-linked SOD1 mutants, even though at least the wild type protein should be competent to acquire zinc in vivo and thereby moderate toxicity.

Second, the primary evidence that the mutants do bind zinc less tightly than wild type SOD1 is the finding that there is accelerated release of the zinc by four mutants (Crow et al. 1997a). However, this release could not be documented under physiological ionic conditions and required the presence of a protein denaturant, raising significant concern for the in vivo relevance. Zinc-depleted mutant subunits were also found to compete less effectively with metal chelators for binding to free zinc (Crow et al. 1997a). Once again, since both protein folding and metal acquisition are facilitated events in vivo, this in vitro measure should not be taken as evidence of the in vivo situation. Indeed, in the only current in vivo test for zinc binding, use of yeast in which the endogenous yeast SOD1 gene was replaced with the human wild type SOD1 or any of a series of ALS-linked mutants has revealed that all mutants examined bind the zinc as effectively as the wild type, as illustrated by their ability to protect yeast from toxicity of environmental zinc (Fig. 3). This finding was confirmed after purification from yeast: all three mutants tested (including SOD1[L38V] used by Estevez et al. 1999) retained normal levels of bound zinc (Goto et al. 2000).

Third, despite at least 67 different disease-related mutations identified in the gene encoding the 153 amino acid SOD1 polypeptide (Andersen et al. 2000),

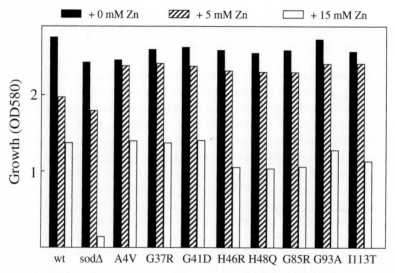

Fig. 3. Familial ALS-SOD1 subunits buffer zinc as effectively as wild type SOD1 in vivo. Growth of *sod1Δ cup1Δ yeast* (JS2004), harboring yeast SOD1-promoted familial ALS- SOD1 mutant expression plasmids (Corson et al. 1998), was monitored in media with varying levels of zinc. Adapted with permission from Williamson et al. (2000)

none of the known mutants lies in any of the four residues that directly coordinate the zinc.

Fourth, in the cell culture model, inhibitors of neuronal NO synthase (nNOS) and immunocytochemical detection of nitrotyrosine, a footprint left by peroxynitrite, were used to demonstrate a dependence of toxicity on NO. Limiting NO production by loss of nNOS would therefore be predicted to ameliorate disease. Yet, in mutant-mediated disease in mice, disruption of the gene for nNOS, accompanied by a 14-fold reduction in detectable NOS activity, does not affect disease onset or progression (Facchinetti et al. 1999).

Fifth, an abundant component of motor neurons is neurofilaments, whose subunits bind zinc in vitro. As proposed by Estevez et al. (1999), competition with neurofilaments would lower mutant SOD1-bound zinc. Therefore, raising the content of neurofilaments would be predicted to exacerbate disease if reduction in zinc bound to mutant SOD1 produces toxicity. However, just the opposite happens. In what represents by far the most robust amelioration of disease to date, raising the content of neurofilaments in motor neuron cell bodies by increasing the synthesis of the NF-H subunit slows disease onset in SOD1^{G37R} mice by six months (Couillard-Despres et al. 1998).

Sixth and finally, SOD1-mediated nitration of neurofilament subunits has been demonstrated in vitro (Crow et al. 1997b). Because the neuron-specific neurofilaments have biological half lives of at least several months and are especially abundant in motor neurons, tyrosine residues on each of the three subunits would be expected to be robust targets for accumulated nitration if peroxynitrite were generated by mutant SOD1. However, use of mass spectrometry to sequence neurofilament subunits isolated from end-stage animals that develop motor neuron disease from expressing either the SOD1^{G37R} or SOD1^{G85R} mutants failed to detect any nitrated peptide with coverage of 100 % of the 20 tyrosine-containing tryptic fragments of NF-L. Nor at any point in the course of the disease in these two animal models could nitrotyrosine be detected on any target protein using a variety of immunologic methods, including the immunocytochemistry approach used by Estevez et al. (1999).

Thus, while the evidence seems firm for peroxynitrite-mediated nitration as a component of cell death arising in vitro from wild type or mutant SOD1 introduced in a zinc-depleted form, available in vivo tests offer no support for this hypothesis reflecting even a portion of the mechanism through which the mutants cause motor neuron disease.

Linking Motor Neuron Growth and Death: Neurofilaments, Axonal Disorganization and Motor Neuron Disease

Neurofilaments, assembled from NF-L (65 kDa), NF-M (95 kDa) and NF-H (115 kDa), are the major intermediate filaments in many types of mature neurons and are particularly abundant in large myelinated axons. All three neurofilament subunits can co-assemble into filaments, although assembly in vivo requires expression of NF-L and a substoichiometric amount of either NF-M or NF-H (Ching and Liem 1993; Lee et al. 1993). Neurofilaments become abundant axonal components only after axons have reached their targets, concomitant with radial growth. Since the initial visualization of neurofilaments by the great neuroanatomists of the 19th century (e.g. Cajal 1903), evidence from several genetic contexts has proven that they are primary determinants of the large increase in axonal diameter that takes place following stable synapse formation. For human motor neurons that elongate with diameters under 1 μm, but achieve final calibers of up to 14 μm, this radial growth phase represents a more than 100-fold increase in cell volume.

In addition to playing a structural role during normal radial growth, abnormal accumulations of neurofilaments are a common early pathological hallmark in patients with sporadic ALS (Carpenter 1968; Chou and Fakadej 1971; Hirano et al. 1984a; reviewed in Hirano 1991) and familial ALS (Hirano et al. 1984b). Two types of morphological changes are associated with this accumulation of filaments apparently early in the degenerative process: first, there are focal accumulation of filaments in cell bodies, often visualized by light microscopy as hyaline inclusions (Schochet et al. 1969; Chou and Fakadej 1971) and neurofilamentous swellings in axons (Carpenter 1968; Cork et al. 1982, 1988; Hirano et al. 1984a); second there are diffuse accumulations of neurofilaments throughout cell bodies, axons and dendrites. This latter type of accumulation is commonly seen in neurons with dispersed endoplasmic reticulum (Cork et al. 1982; Hirano et al. 1984a, b; Wiley et al. 1987), a phenomenon known as chromatolysis. These observations led to the hypothesis that abnormalities in neurofilament organization may be involved in the pathogenesis of ALS. Indeed, recent efforts have demonstrated that disease in the now classic report of familial ALS (Hirano et al. 1984b), in which neurofilament misaccumulation is the prominent pathologic hallmark, arises from $SOD1^{A4V}$ (Shibata et al. 1996a), by far the most abundant SOD1 mutation in North America. Neurofilamentous accumulations have also been found arising from a second mutation $SOD1^{I113T}$ (Rouleau et al. 1996).

To examine directly whether aberrant neurofilament subunits could be a direct cause of motor neuron disease that mimicked the selective killing of motor neurons found in human ALS, several lines of transgenic mice were constructed to express a point mutation (L394P) at the end of the rod domain of NF-L (Lee et al. 1994). Transgenic mice showed abnormal gait, reduced activity and weakness in both upper and lower limbs as early as 18 days after birth. These abnormalities progressively increased in severity and, in most lines, progressed to death. Examination of spinal cord from animals expressing this transgene revealed loss of spinal cord motor neurons and a dramatic loss of motor axons reminiscent of ALS (Lee et al. 1994). As in many ALS cases, the surviving ventral horn motor neurons were chromatolytic and were grossly distended with material that stained lightly with toluidine blue, but stained intensely with silver (Lee et al. 1994). Moreover, as reported frequently for ALS (see, for example, Carpenter 1968; Hirano et al. 1984a, b), neurofilament-rich swellings of proximal axons were frequently found in the initial unmyelinated segment and in the root exit zones.

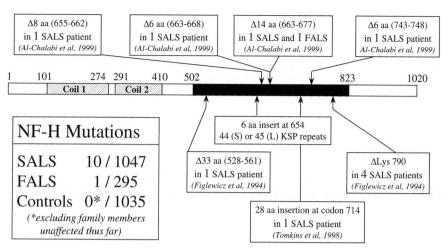

Fig. 4. Mutations in sporadic (SALS) and familial ALS (FALS) found in the gene encoding the large neurofilament subunit NF-H. Reprinted with permission from Cleveland (1999)

The evidence that, in mice, mutant neurofilaments could directly provoke motor neuron disease and the accompanying neurofilamentous misaccumulation fueled efforts to test whether neurofilament mutations underlay disease. Early efforts focused on the 80 % of familial ALS not arising from SOD1 mutation, but did not identify neurofilament mutants linked to inherited disease (Vechio et al. 1996; Rooke et al. 1996). Subsequent efforts (Figlewicz et al. 1994; Tomkins et al. 1998, Al-Chalabi et al. 1999) examining the repetitive tail domain of the large neurofilament subunit NF-H, especially the extensive study of Al-Chalabi et al. (1998), have now identified a set of small in-frame deletions or insertions in ~1 % of more than 1300 ALS patients, almost all of which appear in "sporadic" cases. Excluding family members not yet affected, no similar mutants have been seen in a comparable number of control DNAs. Search of this one domain alone has thus yielded mutations in the overall patient population about half as frequent as SOD1 mutants (Fig. 4). It should be realized that variation in timing of disease onset and incomplete penetrance can befuddle identifying an underlying genetic component. While the known neurofilament sequence variants are surely not by themselves capable of provoking disease with high penetrance, the collective evidence now strongly hints that variants in neurofilaments are, at the least, important risk factors for apparently sporadic disease. Relevant here is the discovery of a mutation in NF-L (Q333P) as the probable cause of one example of inherited Charcot-Marie-Tooth disease (Mersiyanova et al. 2000), a motor neuron disorder characterized by axonal degeneration.

A pair of additional genetic experiments in mice has demonstrated that neurofilaments do participate in disease initially provoked by mutant SOD1. By using gene disruption to eliminate neurofilaments from mice expressing the SOD1^{G85R} mutant, onset and progression of disease have been found to be slowed by 40 days, despite the significant initial disadvantage of loss of $\sim$ 15 % of the motor neurons early in postnatal life arising from the absence of neurofilaments (Williamson et al. 1998). Even more provocative, decreasing the axonal amount of neurofilaments by trapping most in the cell bodies through transgene-increased expression of the large neurofilament subunit NF-H extends life span in SOD1^{G37R} mice by six months (a 65 % increase in longevity; Couillard-Despres et al. 1998). While several other strategies have successfully slowed disease in SOD1 mice, this last neurofilament strategy represents by far the most striking slowing of disease.

Deficits in Axonal Transport Are Early Features of SOD1-Mediated ALS

One possible mechanism for the pathogenesis of ALS (especially one influenced by neurofilaments) is a defect that either directly or indirectly impairs slow axonal transport, leading to aberrant accumulation of neurofilaments in the proximal axon and inhibiting further transport of neurofilaments and potentially other substrates. In view of this, degeneration may arise from the simple mechanism of axonal strangulation. Transport is comprised of two components: movement of vesicles and mitochondria by kinesin and related proteins (fast transport) and movement of the major structural components of the neuron, many enzymes and other cytoplasmic proteins (slow transport). Slow transport can be divided into two components based on rate of movement: SCa ($\sim$ 0.5 mm/day) containing neurofilament proteins, tubulin and actin, and SCb ($\sim$1–2 mm/day), containing tubulin, actin and other cytoplasmic proteins. The mechanism(s) of slow transport (and motor(s) that power it) is not known.

Support for influence of neurofilaments on slow transport came from transgenic mice expressing wild type human NF-H mRNA at levels two-fold over endogenous murine mRNA levels. These animals have perikaryal and axonal accumulation of neurofilaments in neurons, accompanied by development of overt clinical abnormalities beginning as early as three to four months of age and characterized by fine tremors, forelimb and muscle atrophy, along with slowing of slow axonal transport (Cote et al. 1993; Collard et al. 1995). Transgenic mice expressing murine NF-H (mNF-H) show similar increases in NF-H expression, cytoskeletal abnormalities and decreased axonal neurofilament transport (Marszalek et al. 1996).

Chronic deficits in transport of selective cargoes of slow transport have now been seen as very early features, initiating up to five months prior to clinical disease onset in mice expressing SOD1^{G85R} and SOD1^{G37R} mutants (Williamson and Cleveland 1999). Deficits in fast transport have been seen prior to disease onset in SOD1^{G93A} mice (Warita et al. 1999). Thus, whether arising from direct damage to the machinery or cargoes of transport, a common deficit from SOD1 mutants that provoke different pathologies is chronically compromised delivery of selected transport cargoes. Since neurofilaments are a major cargo whose presence slows the speed of slow transport, this offers at least a partial explanation for the vulnerability of the large motor axons at risk. The motor neurons, with large, neurofilament-rich axons that in humans are up to a meter in length, are among the body's longest and largest cells (up to 5000 times the volume of a typical cell), almost all components of which must be transported into and along the axon.

Aggregates of SOD1 Are a Common Feature of the ALS-Linked SOD1 Mutants

Studies of several neurodegenerative diseases (Alzheimer's disease, prion diseases, triplet repeat expansion diseases such as Huntington's) have revealed a common feature: protein aggregates. This commonality has fueled a long-standing debate over whether these aggregates are key to the pathogenesis, harmless byproducts, or potentially beneficial through sequestration of aberrant products. The possibility that aggregates might be contributors to ALS was significantly boosted by the finding that all examples of SOD1-mutant mediated disease in mice develop prominent, cytoplasmic, intracellular inclusions in motor neurons, and in some cases within the astrocytes surrounding them as well. These aggregates develop by the onset of clinical disease, in some cases (e.g., SOD1^{G85R} mice) representing the first pathologic sign of disease, and later increase markedly in abundance during disease progression (Bruijn et al. 1997b).

Aggregation, or at least misfolded subunits, is thus apparently a characteristic of the SOD1 mutants. This has been reproduced selectively in motor neurons by microinjection of genes encoding mutant SOD1s, but not wild type. This approach has yielded what may be the best in vitro model of disease, including aggregates selectively in motor (but not sensory or hippocampal) neurons (Durham et al. 1997), followed by cell death that can be slowed by apoptosis inhibitors. Aspects of toxicity may arise either through aberrant chemistry mediated by the misfolded aggregated mutants or through loss by co-precipitation of an essential component or components, for example, by saturating the protein degradation machinery and/or protein folding chaperones. Consistent with the latter, increasing the level of the stress-inducible chaperone HSP 70 reduced mutant driven aggregates and ameliorated toxicity in this in vitro cell model (Bruening et al. 1999).

Future Prospects: Testing the Oxidative and Aggregation Hypotheses. Will Removing the Catalytic Copper from Mutant SOD1 Ameliorate or Eliminate Toxicity?

Despite the evidence against the current proposed substrates as sources of a toxicity common to the various SOD1 mutants, no conclusion can yet be drawn as to whether, or to what degree, oxidative damage plays a role in disease. There may, of course, be yet unidentified, alternative substrates that initiate a cascade of oxidative damage catalyzed by mutant SOD1. Whatever the case, the seminal discovery that copper acquisition by SOD1 in yeast requires CCS, a specific Copper

Chaperone for SOD1 (Culotta et al. 1997), has allowed design of a test, in a single experiment, of all possible, copper-mediated toxicities. Since both the human wild type and mutant SOD1s apparently load copper in vivo through the action of a mammalian CCS (Corson et al. 1998; Wong et al. 2000), production of SOD1 mutant-expressing mice that are also deleted for the single mammalian CCS will directly test the requirement for copper (and any catalyzed oxidative damage) in SOD1 mutant toxicity. This experiment, which has the power to disprove copper-mediated oxidative mechanisms should toxicity prove insensitive to copper loading, is well underway in at least one group, with a clear answer anticipated by the end of 2000. It should be noted that the outcome of the test of the influence of copper loading on toxicity will probably also bear directly on distinguishing the potential contributions of copper catalysis versus SOD1 aggregation.

As to the aggregation hypothesis, direct efforts to reduce aggregates and their potential role in disease are in progress in at least two groups, by the introduction of transgenes encoding elevated levels of one or more protein folding chaperones like Hsp 70 (e.g., the Hsp 70-overexpressing mice of Marber et al. 1995). Combined with the CCS deletion mating, this pair of transgenic approaches may in the next several months confirm, refute, or at least refine, proposals for the mechanism of selective killing of motor neurons in ALS.

References

Abe K, Pan LH, Watanabe M, Kato T, Itoyama Y (1995) Induction of nitrotyrosine-like immunoreactivity in the lower motor neurons of amyotrophic lateral sclerosis. Neurosci Lett 199: 152–154

Abe K, Pan LH, Watanabe M, Kato T, Itoyama Y (1997) Upregulation of protein-tyrosine nitration in the anterior horn cells of amyotrophic lateral sclerosis. Neurol Res 19: 124–128

Al-Chalabi A, Andersen PM, Nilsson P, Chioza B, Andersson JL, Russ C, Shaw CE, Powell JF, Leigh PN (1999) Deletions of the heavy neurofilament subunit tail in amyotrophic lateral sclerosis. Human Mol Genet 8: 157–164

Andersen PM, Morita M, Brown RH Jr. (2000) Genetics of amyotrophic lateral sclerosis: an overview, pp. 223–250. In: Brown RH Jr., Meininger V, Swash M (eds) Amyotrophic lateral sclerosis. Martin Dunitz, London

Beckman JS, Carson M, Smith CD, Koppenol WH (1993) ALS, SOD and peroxynitrite. Nature 364: 584

Beal MFR, Ferrante J, Browne SE, Matthews RT, Kowall NW, Brown RH Jr (1997) Increased 3-nitrotyrosine in both sporadic and familial amyotrophic lateral sclerosis. Ann Neurol 42: 644–654

Borchelt DR, Lee MK, Slunt HS, Guarnieri M, Xu ZS, Wong PC, Brown RH Jr, Price DL, Sisodia SS, Cleveland DW (1994) Superoxide dismutase 1 with mutations linked to familial amyotrophic lateral sclerosis possesses significant activity. Proc Natl Acad Sci USA 91: 8292–8296

Borchelt DR, Guarnieri M, Wong PC, Lee MK, Slunt HS, Xu ZS, Sisodia SS, Price DL, Cleveland DW (1995) Superoxide dismutase 1 subunits with mutations linked to familial amyotrophic lateral sclerosis do not affect wild-type subunit function. J Biol Chem 270: 3234–3238

Bowling AC, Schulz JB, Brown RH Jr, Beal MF (1993) Superoxide dismutase activity, oxidative damage, and mitochondrial energy metabolism in familial and sporadic amyotrophic lateral sclerosis. J Neurochem 61: 2322–2325

Bruening W, Roy J, Giasson B, Figlewicz DA, Mushynski WE, Durham HD (1999) Up-regulation of protein chaperones preserves viability of cells expressing toxic Cu/Zn-superoxide dismutase mutants associated with amyotrophic lateral sclerosis. J Neurochem 72: 693–699

Bruijn LI, Beal MF, Becher MW, Schulz JB, Wong PC, Price DL, Cleveland DW (1997a) Elevated free nitrotyrosine levels, but not protein-bound nitrotyrosine or hydroxyl radicals, throughout amyotrophic lateral sclerosis (ALS)-like disease implicate tyrosine nitration as an aberrant in vivo property of one familial ALS-linked superoxide dismutase 1 mutant. Proc Natl Acad Sci USA 94: 7606–7611

Bruijn LI, Becher MW, Lee MK, Anderson KL, Jenkins NA, Copeland NG, Sisodia SS, Rothstein JD, Borchelt DR, Price DL, Cleveland DW (1997b) ALS-linked SOD1 mutant SOD1^{G85R} mediates damage to astrocytes and promotes rapidly progressive disease with SOD1-containing inclusions. Neuron 18: 327–338

Bruijn LI, Houseweart MK, Kato S, Anderson KL, Anderson SD, Ohama E, Reaume AG, Scott RW, Cleveland DW (1998) Aggregation and motor neuron toxicity of an ALS-linked SOD1 mutant independent from wild-type SOD1. Science 281: 1851–1854

Cajal SR (1903) Embryogenesis of neurofibrils. Trab Lab Invest Biol Univ Madrid 2: 219–225

Carpenter S (1968) Proximal axonal enlargement in motor neuron disease. Neurology 18: 841–851

Ching GY, Liem RK (1993) Assembly of type IV neuronal intermediate filaments in nonneuronal cells in the absence of preexisting cytoplasmic intermediate filaments. J Cell Biol 122: 1323–1335

Chou SM, Fakadej AV (1971) Ultrastructure of chromatolytic motor neurons and anterior spinal roots in a case of Werdnig-Hoffman disease. J Neuropathol Exp Neurol 30: 42–55

Chou SM, Wang HS, Taniguchi A (1996) Role of SOD-1 and nitric oxide/cyclic GMP cascade on neurofilament aggregation in ALS/MND. J Neurol Sci 139 (Suppl): 16–26

Cleveland DW (1999) From Charcot to SOD1: mechanisms of selective motor neuron death in ALS. Neuron 24: 515–520

Cleveland DW, Laing N, Hurse PV, Brown RH (1995) Toxic mutants in Charcot's sclerosis. Nature 378: 342–343

Collard JF, Cote F, Julien JP (1995) Defective axonal transport in a transgenic mouse model of amyotrophic lateral sclerosis. Nature 375: 61–64

Cork LC, Griffin JW, Choy C, Padula CA, Price DL (1982) Pathology of motor neurons in accelerated hereditary canine spinal muscular atrophy. Lab Invest 46: 89–99

Cork LC, Troncoso JC, Klavano GG, Johnson ES, Sternberger LA, Sternberger NH, Price DL (1988) Neurofilamentous abnormalities in motor neurons in spontaneously occurring animal disorders. J Neuropathol Exp Neurol 47: 420–431

Corson LB, Strain JJ, Culotta VC, Cleveland DW (1998) Chaperone-facilitated copper binding is a property common to several classes of familial amyotrophic lateral sclerosis-linked superoxide dismutase mutants. Proc Natl Acad Sci USA 95: 6361–6366

Cote F, Collard JF, Julien JP (1993) Progressive neuronopathy in transgenic mice expressing the human neurofilament heavy gene: a mouse model of amyotrophic lateral sclerosis. Cell 73: 35–46

Couillard-Despres S, Zhu Q, Wong PC, Price DL, Cleveland DW, Julien JP (1998) Protective effect of neurofilament heavy gene overexpression in motor neuron disease induced by mutant superoxide dismutase. Proc Natl Acad Sci USA 95: 9626–9630

Crow JP, Sampson JB, Zhuang Y, Thompson JA, Beckman JS (1997a) Decreased zinc affinity of amyotrophic lateral sclerosis-associated superoxide dismutase mutants leads to enhanced catalysis of tyrosine nitration by peroxynitrite. J Neurochem 69: 1936–1944

Crow JP, Ye YZ, Strong M, Kirk M, Barnes S, Beckman JS (1997b) Superoxide dismutase catalyzes nitration of tyrosines by peroxynitrite in the rod and head domains of neurofilament L. J Neurochem 69: 1945–1953

Culotta VC, Klomp LW, Strain J, Casareno RL, Krems B, Gitlin JD (1997) The copper chaperone for superoxide dismutase. J Biol Chem 272: 23469–23472

Dal Canto MC, Gurney ME (1994) Development of central nervous system pathology in a murine transgenic model of human amyotrophic lateral sclerosis. Am J Pathol 145: 1271–1279

Dal Canto MC, Gurney ME (1995) Neuropathological changes in two lines of mice carrying a transgene for mutant human Cu,Zn SOD, and in mice overexpressing wild type human SOD: a model of familial amyotrophic lateral sclerosis (FALS). Brain Res 676: 25–40

Dal Canto MC, Gurney ME (1997) A low expressor line of transgenic mice carrying a mutant human Cu,Zn superoxide dismutase (SOD1) gene develops pathological changes that most closely resemble those in human amyotrophic lateral sclerosis. Acta Neuropathol 93: 537–550

Deng HX, Hentati A, Tainer JA, Iqbal Z, Cayabyab A, Hung WY, Getzoff ED, Hu P, Herzfeldt B, Roos RP, Warner C, Siddique T (1993) Amyotrophic lateral sclerosis and structural defects in Cu,Zn superoxide dismutase. Science 261: 1047–1051

Durham HD, Roy J, Dong L, Figlewicz DA (1997) Aggregation of mutant Cu/Zn superoxide dismutase proteins in a culture model of ALS. J Neuropathol Exp Neurol 56: 523–530

Estevez AG, Crow JP, Sampson JB, Reiter C, Zhuang Y, Richardson GJ, Tarpey MM, Barbeito L, Beckman JS (1999) Induction of nitric oxide-dependent apoptosis in motor neurons by zinc-deficient superoxide dismutase. Science 286: 2498–2500

Facchinetti F, Sasaki M, Cutting FB, Zhai P, MacDonald JE, Reif D, Beal MF, Huang PL, Dawson TM, Gurney ME, Dawson VL (1999) Lack of involvement of neuronal nitric oxide synthase in the pathogenesis of a transgenic mouse model of familial amyotrophic lateral sclerosis. Neuroscience 90: 1483–1492

Ferrante RJ, Shinobu LA, Schulz JB, Matthews RT, Thomas CE, Kowall NW, Gurney ME, Beal MF (1997) Increased 3-nitrotyrosine and oxidative damage in mice with a human copper/zinc superoxide dismutase mutation. Ann Neurol 42: 326–334

Figlewicz DA, Krizus A, Martinoli MG, Meininger V, Dib M, Rouleau GA, Julien JP (1994) Variants of the heavy neurofilament subunit are associated with the development of amyotrophic lateral sclerosis. Human Mol Genet 3: 1757–1761

Fridovich I (1995) Superoxide radical and superoxide dismutases. Annu Rev Biochem 64: 97–112

Gonatas NK, Stieber A, Mourelatos Z, Chen Y, Gonatas JO, Appel SH, Hays AP, Hickey WF, Hauw JJ (1992) Fragmentation of the Golgi apparatus of motor neurons in amyotrophic lateral sclerosis. Am J Pathol 140: 731–737

Goto JJ, Zhu H, Sanchez RJ, Nersissian A, Gralla EB, Valentine JS, Cabelli DE (2000) Loss of in vitro metal ion binding specificity in mutant copper-zinc superoxide dismutases associated with familial amyotrophic lateral sclerosis. J Biol Chem 275: 1007–1014

Gurney ME, Pu H, Chiu AY, Dal Canto MC, Polchow CY, Alexander DD, Caliendo J, Hentati A, Kwon YW, Deng HX (1994) Motor neuron degeneration in mice that express a human Cu,Zn superoxide dismutase mutation. Science 264: 1772–1775

Halliwell B (1994) Free radicals, antioxidants, and human disease: curiosity, cause, or consequence? Lancet 344: 721–724

Hirano A (1991) Cytopathology of amyotrophic lateral sclerosis. In: Rowland LP (ed) Advances in neurology. Volume 56. Amyotrophic lateral sclerosis and other motor neuron diseases. New York; Raven Press, pp 91–101

Hirano A, Donnenfeld H, Sasaki S, Nakano I (1984a) Fine structural observations of neurofilamentous changes in amyotrophic lateral sclerosis. J Neuropathol Exp Neurol 43: 461–470

Hirano A, Nakano I, Kurland LT, Mulder DW, Holley PW, Saccomanno G (1984b) Fine structural study of neurofibrillary changes in a family with amyotrophic lateral sclerosis. J Neuropathol Exp Neurol 43: 471–480

Kato S, Shimono M, Watanabe Y, Nakashima K, Takahashi K, Ohama E (1996) Familial amyotrophic lateral sclerosis with a two base pair deletion in superoxide dismutase 1 gene: multisystem degeneration with intracytoplasmic hyaline inclusions in astrocytes. J Neuropathol Exp Neurol 55. 1089–1101

Kato S, Hayashi H, Nakashima K, Nanba E, Kato M, Hirano A, Nakano I, Asayama K, Ohama E (1997) Pathological characterization of astrocytic hyaline inclusions in familial amyotrophic lateral sclerosis. Am J Pathol 151: 611–620

Lee MK, Xu Z, Wong PC, Cleveland DW (1993) Neurofilaments are obligate heteropolymers in vivo. J Cell Biol 122: 1337–1350

Lee MK, Marszalek JR, Cleveland DW (1994) A mutant neurofilament subunit causes massive, selective motor neuron death: Implications for the pathogenesis of human motor neuron disease. Neuron 13: 975–988

Leigh PN, Whitwell H, Garofalo O, Buller J, Swash M, Martin JE, Gallo JM, Weller RO, Anderton BH (1991) Ubiquitin-immunoreactive intraneuronal inclusions in amyotrophic lateral sclerosis. Morphology, distribution, and specificity. Brain 114: 775–788

Liu XF, Elashvili I, Gralla EB, Valentine JS, Lapinskas P, Culotta VC (1992) Yeast lacking superoxide dismutase. Isolation of genetic suppressors. J Biol Chem 267: 18298–18302

Marber MS, Mestril R, Chi SH, Sayen MR, Yellon DM, Dillmann WH (1995) Overexpression of the rat inducible 70-kD heat stress protein in a transgenic mouse increases the resistance of the heart to ischemic injury. J Clin Invest 95: 1446–1456

Marszalek JR, Williamson TL, Lee MK, Xu Z-S, Crawford TO, Hoffman PN, Cleveland DW (1996) Neurofilament subunit NF-H modulates axonal diameter by affecting the rate or neurofilament transport. J Cell Biol 135: 711–724

McNamara JO, Fridovich I (1993) Human genetics. Did radicals strike Lou Gehrig? Nature 362: 20–21

Mersiyanova IV, Perepelov AV, Polyakov AV, Sitnikov VF, Dadali EL, Oparin RB, Petrin AN, Evgrafov OV (2000) A new variant of Charcot-Marie-Tooth disease type 2 is probably the result of a mutation in the neurofilament-light gene. Am J Human Genet 67: 37–46

Orrell R, de Belleroche J, Marklund S, Bowe F, Hallewell R (1995) A novel SOD mutant and ALS. Nature 374: 504–505

Peled-Kamar M, Lotem J, Wirguin I, Weiner L, Hermalin A, Groner Y (1997) Oxidative stress mediates impairment of muscle function in transgenic mice with elevated level of wild-type Cu/Zn superoxide dismutase. Proc Natl Acad Sci USA 94: 3883–3887

Reaume AB, Elliott JL, Hoffman EK, Kowall NW, Ferrante RJ, Siwek DF, Wilcox HM, Flood DG, Beal MF, Brown RH, Scott RW, Snider WD (1996) Motor neurons in Cu/Zn superoxide dismutase-deficient mice develop normally but exhibit enhanced cell death after axonal injury. Nature Genet 13: 43–47

Ripps ME, Huntley GW, Hof PR, Morrison JH, Gordon JW (1995) Transgenic mice expressing an altered murine superoxide dismutase gene provide an animal model of amyotrophic lateral sclerosis. Proc Natl Acad Sci USA 92: 689–693

Robberecht W, Sapp P, Viaene MK; Rosen D, McKenna-Yasek D, Haines J, Horvitz R, Theys P, Brown R (1994) Cu/Zn superoxide dismutase activity in familial and sporadic amyotrophic lateral sclerosis. J Neurochem 62: 384–387

Rooke K, Figlewicz DA, Han F, Rouleau GA (1996) Analysis of the KSP repeat of the neurofilament heavy subunit in familial amyotrophic lateral sclerosis. Ann Neurol 46: 789–790

Rosen DR, Siddique T, Patterson D, Figlewicz DA, Sapp P, Hentati A, Donaldson D, Goto J, O'Regan JP, Deng HX, Brown R (1993) Mutations in Cu/Zn superoxide dismutase gene are associated with familial amyotropic lateral sclerosis. Nature 362: 59–62

Rouleau GA, Clark AW, Rooke K, Pramatarova A, Krizus A, Suchowersky O, Julien J-P, Figlewicz D (1996) SOD1 mutation is associated with accumulation of neurofilaments in amyotrophic lateral sclerosis. Ann Neurol 39: 128–131

Schochet SS, Hardman JM, Ladewig PP, Earle KM (1969) Intraneuronal conglomerates in sporadic motor neuron disease. A light and electron microscopic study. Arch Neurol 20: 548–553

Shefner JM, Reaume AG, Flood DG, Scott RW, Kowall NW, Ferrante RJ, Siwek DF, Upton-Rice M, Brown RH Jr (1999) Mice lacking cytosolic copper/zinc superoxide dismutase display a distinctive motor axonopathy. Neurology 53: 1239–1246

Shibata N, Hirano A, Kobayashi M, Siddique T, Deng H-X, Hung W-Y, Kato T, Asayama K (1996) Intense superoxide dismutase-1 immunoreactivity in intracytoplasmic hyaline inclusions of familial amyotrophic lateral sclerosis with posterior column involvement. J Neuropathol Exp Neurol 55: 481–490

Siddique T, Deng HX (1996) Genetics of amyotrophic lateral sclerosis. Human Mol Genet 5 (Spec No): 1465–70

Singh RJ, Karoui H, Gunther MR, Beckman JS, Mason RP, Kalyanaraman B (1998) Reexamination of the mechanism of hydroxyl radical adducts formed from the reaction between familial amyotrophic lateral sclerosis-associated Cu,Zn superoxide dismutase mutants and H2O2. Proc Natl Acad Sci USA 95: 6675–6680

Tomkins J, Usher P, Slade JY, Ince PG, Curtis A, Bushby K, Shaw PJ (1998) Novel insertion in the KSP region of the neurofilament heavy gene in amyotrophic lateral sclerosis (ALS). Neuroreport 9: 3967–3970

Troy CM, Derossi S, Prochiantz A, Greene LA, Shelanski ML (1996) Down-regulation of SOD1 leads to cell death by the NO-peroxynitrite pathway. J Neurosci 16: 253–261

Tu PH, Raju P, Robinson KA, Gurney ME, Trojanowski JQ, Lee VM-Y (1996) Transgenic mice carrying a human mutant superoxide dismutase transgene develop neuronal cytoskeletal pathology resembling human amyotrophic lateral sclerosis lesions. Proc Natl Acad Sci USA 93: 3155–3160

Vechio JD, Bruijn LI, Xu Z, Brown RH Jr, Cleveland DW (1996) Sequence variants in human neurofilament proteins: absence of linkage to familial amyotrophic lateral sclerosis. Ann Neurol 40: 603–610

Warita H, Itoyama Y, Abe K (1999) Selective impairment of fast anterograde axonal transport in the peripheral nerves of asymptomatic transgenic mice with a G93A mutant SOD1 gene. Brain Res 819: 120–131

Wiedau-Pazos M, Goto JJ, Rabizadeh S, Gralla EB, Roe JA, Lee MK, Valentine JS, Bredesen DE (1996) Altered reactivity of superoxide dismutase in familial amyotrophic lateral sclerosis. Science 271: 515–518

Wiley CA, Love S, Skoglund RR, Lampert PW (1987) Infantile neurodegenerative disease with neuronal accumulation of phosphorylated neurofilaments. Acta Neuropathol (Berl) 72: 369–376

Williamson TL, Cleveland DW (1999) Slowing of axonal transport is a very early event in the toxicity of ALS-linked SOD1 mutants to motor neurons. Nature Neurosci 2. 50–56

Williamson-TL, Bruijn LI, Zhu Q, Anderson KL, Anderson SD, Julien JP, Cleveland DW (1998) Absence of neurofilaments reduces the selective vulnerability of motor neurons and slows disease caused by a familial amyotrophic lateral sclerosis-linked superoxide dismutase 1 mutant. Proc Natl Acad Sci USA 95: 9631–9636

Williamson TL, Corson LB, Huang A, Burlingame J, Liu L, Bruijn I, Cleveland DW (2000) Toxicity of ALS-linked SOD1 mutants. Science 288: 399a–400a

Wong PC, Pardo CA, Borchelt DR, Lee MK, Copeland NG, Jenkins NA, Sisodia SS, Cleveland DW, Price DL (1995) An adverse property of a familial ALS-linked SOD1 mutation causes motor neuron disease characterized by vacuolar degeneration of mitochondria. Neuron 14: 1105–1116

Wong PC, Waggoner D, Subramaniam JR, Tessarollo L, Bartnikas TB, Culotta VC, Price DL, Rothstein J, Gitlin JD (2000) Copper chaperone for superoxide dismutase is essential to activate mammalian Cu/Zn superoxide dismutase. Proc Natl Acad Sci USA 97: 2886–2891

Yu BP (1994) Cellular defenses against damage from reaactive oxygen species. Physiol Rev 74: 138–162

Pathological Mechanisms in Huntington's Disease and Other Polyglutamine Expansion Diseases

A. Lunkes, G. Yvert, Y. Trottier, D. Devys, and J. L. Mandel

Introduction

Since 1991, nine monogenic neurodegenerative diseases have been shown to be caused by moderate expansion of a CAG repeat coding for a polyglutamine stretch in specific target proteins. These disorders include Huntington's disease (HD) and various spinocerebellar ataxias (SCAs), and their pathogenic mechanism has been the object of intense recent studies. While it is clear that the mutations confer a gain of toxic property to the target proteins, correlated with appearance of a common pathologic epitope and of self aggregation properties, much remains to be learned concerning how these elongated polyglutamines cause neuronal dysfunction and death, and what features can account for the selectivity with which various neuronal populations are affected. Important observations have recently been made, notably by use of cellular or transgenic animal models, that give new insight into the mechanisms of polyglutamine diseases and raise hopes for therapies.

Huntington's Disease: Clinical Features

Huntington's disease (HD) is the most frequent and the best known of the polyglutamine expansion diseases. It has a prevalence of about one in 12,000–15,000 in Western European populations (Harper 1992). The diseases begins gradually, leading to mood disturbances, involuntary movements (chorea) and cognitive impairment, and progresses inexorably to death within 15–20 years after onset (Harper 1991). The motor disability affects both involuntary and voluntary movements. Chorea is observed in ≥90% of patients, and dystonia becomes a prominent feature in later stages of the illness. A progressive loss of coordination of voluntary movements leads to a state in which voluntary movement is impossible. The cognitive disorder begins with loss of mental flexibility and slowing of intellectual processes, and progresses to profound dementia. Mood abnormalities often appear a few years before the onset of the movement disorder. Age at onset of symptoms is classically between 30 and 50 years; however, juvenile and late-onset cases also occur. The juvenile forms (onset before the age of 20) represent about 6–10% of cases and are almost exclusively transmitted by the father. The disease is dominantly inherited, with almost complete penetrance (see

Research and Perspectives in Alzheimer's Diseases
Beyreuther/Christen/Masters (Eds.)
Neurodegenerative Disorders
© Springer-Verlag Berlin Heidelberg 2001

below), and homozygous or sporadic cases are rare. The fact that the rare homozygous patients do not show increased severity provided a decisive argument in
favor of a "gain of function" conferred by the mutation (as dominant diseases
caused by a loss of function mutation are much more severe in the homozygous
state; Gusella et al. 1996).

Huntington's Disease: Neuropathology

The clinical progression of HD is paralleled by brain degeneration. While caudate
and putamen show the most dramatic change, there is an overall atrophy of the
brain. In advanced cases, the brain often weighs 20–25 % less than normal.
Within the striatum, there is selective loss of medium spiny GABAergic neurons,
but a relative preservation of interneurons. The overall architecture of the caudate nucleus and the adjacent putamen is destroyed, and gliosis may be prominent. Prior to cell death, neuronal dysfunction is manifested by anomalies of dendritic endings and changes in spine density and shape. Extensive neuronal loss
also occurs in the deep layers of the cerebral cortex, and may affect other brain
structures (thalamus, hypothalamic lateral nucleus, etc.).

The Expansion Mutation and Genotype:Phenotype Correlations

An expansion of a CAG repeat in exon 1 of the HD gene is the exclusive genetic
defect causing HD (The Huntington's Disease Collaborative Research Group
1993). The HD gene encodes a very large protein (350 kDa) of unknown function,
named huntingtin. The CAG repeat in the first exon is polymorphic in the normal population, varying from 6 to 35 units. It codes for a polyglutamine stretch
in the amino-terminus of huntingtin. In HD patients, the CAG repeat is expanded
from 36 to 120 repeats. There is an inverse correlation between the length of the
expansion and the age of onset (the repeat size of the mutated allele accounts for
about 70 % of the variance of age of onset; Brinkman et al. 1997; Gusella et al.
1996; Rubinsztein et al. 1997). Most mutated alleles contain between 40 and 55
CAGs and are associated with middle age of onset (Gusella et al. 1996). Between
36 and 39 repeats, penetrance of the disease is not complete, and some carriers of
such alleles may thus remain healthy until ≥80 years of age (Brinkman et al.
1997). Patients with 60 repeats and more present a juvenile onset. Although the
correlation between the length of the expansion and the age of onset is well
established, the wide confidence interval renders this correlation of little use for
individual prediction of age of onset. The identification of other putative modi-
fier genes or environmental factors that may account for some of the variation of
age of onset would be of considerable interest, as they could provide additional
targets for therapeutic interventions (Rubinsztein et al. 1997).
   The mutated alleles show a slight instability in somatic tissues of patients. In
contrast, a strong instability is observed in male germ cells (sperm), in which the
size of the CAG repeat is often very heterogenous and larger than the size mea-

sured in blood of the same patient. This accounts for the paternal bias of trans-mission of the juvenile forms. In such a case, the father transmits a mutated HD allele to his child with an increase in length of ≥ 10 CAGs, leading to anticipation in the age of onset. However, in most of paternal transmission, allelic instability is rather limited and will have thus a limited impact on the child's age of onset. In maternal transmissions, the mutated alleles remain quite stable. Sporadic cases are rare; they may occur through new germline mutations, arising (at low fre-quency) from males carrying an allele in the high normal range (30–35 CAGs).

Huntingtin is a cytoplasmic protein found in many peripheral tissues but is particularly abundant in neurons. Its wide expression in brain contrasts with the selectivity of neuronal loss in HD (Trottier et al. 1995a). In brain, huntingtin localizes in part in nerve endings, suggesting a role in synaptic vesicle transport. While its exact function remains unknown, huntingtin has an important role during early embryonic development. Mouse embryos homozygous for inactivat-ing mutations of the HD gene cannot complete gastrulation, and mice with a strong reduction of HD gene expression show abnormal brain development (White et al. 1997).

Polyglutamine Expansion in Other Neurodegenerative Disorders

The class of neurodegenerative s caused by an expansion of a CAG/polyglutamine repeat has grown steadily since 1991. The last to be added to the list is a type of spinocerebellar ataxia, named SCA7, which is the only one of these diseases that also affects retina (David et al. 1997). More recently, a single case of polygluta-mine expansion arising as a new mutation (from 39 gln to 63 gln) affecting the TATA-binding protein (TBP, a basal transcription factor) has been described in a patient with very severe cerebellar atrophy (Koide et al. 1999). As for HD, all these diseases show a strong inverse correlation between age of onset of clinical symptoms and the length of the abnormal repeat (Gusella et al. 1996). The patho-logical threshold is strikingly similar to HD in four of these diseases (from 35 to 40 gln; Table 1). Below this threshold, the high degree of polymorphism for the CAG/gln repeat observed in the normal population indicates that these proteins tolerate a wide variation in polyglutamine length without adverse effect. The nine diseases are characterized by slowly progressing neuronal degeneration in selected regions of the brain that differ (with some overlap) between the various diseases (Table 1). This selectivity of neuronal death does not show obvious cor-relation with the rather ubiquitous pattern of expression of the cognate proteins (with the exception of SCA6; Zoghbi 1997). The function of these proteins is unknown in six of the nine cases, with the exceptions being the androgen recep-tor implicated in spinal and bulbar muscular atrophy (SBMA or Kennedy dis-ease), a calcium channel subunit mutated in SCA6, and the single case of muta-tion in TBP. The proteins do not resemble each other, except for the polygluta-mine tract.

The genetic features of SBMA (the first identified polyglutamine expansion disease) indicated that, as for HD, the expansion mutation causes a gain of toxic

Table 1. Polyglutamine disorders

Disease	Sites of neuropathology	Repeat Number Normal	Disease	Intracellular localization of encoded protein
SBMA	Motor neurons (anterior horn cells, bulbar neurons) and dorsal root ganglia	11–34	40–62	Nuclear (androgen receptor)
DRPLA	Globus pallidus, dentatorubral and subthalamic nucleus	7–35	49–88	Cytoplasmic
HD	Striatum (medium spiny neurons) and cortex in late stage	6–35	36–121	Cytoplasmic
SCA1	Cerebellar cortex (Purkinje cells), dentate nucleus and brain stem	6–39	40–81	Nuclear, cytoplasmic
SCA2	Cerebellum, pontine nuclei, subtantia nigra	15–29	35–64	Cytoplasmic
SCA3	Substantia nigra, globus pallidus, pontine nucleus, cerebellar cortex	13–42	61–84	Cytoplasmic
SCA6	Cerebellar and mild brain stem atrophy	4–18	21–30	Transmembrane (calcium channel subunit αIA)
SCA7	Photoreceptors and bipolar cells, cerebellar cortex, brainstem	7–17	37–300	Nuclear, cytoplasmic
"TBP"	Severe cerebellar atrophy	25–42	63 (single case)	Nuclear (basal transcription factor)

property and not a loss of function of the cognate gene or protein (mutations that cause a complete loss of function of the androgen receptor gene are responsible for the testicular feminization syndrome, with no neurological symptomatology). By analogy, the same gain of toxic property is expected to also occur in the other diseases, with the possible exception of SCA6. The latter disease is characterized by a much lower pathological threshold of polyglutamine size, and, most importantly, conventional loss of function mutations in the same target gene result in acetazolamide-responsive episodic ataxia, a disease that shares overlapping clinical features with SCA6 and that leads to progressive cerebellar degeneration in some patients (Zoghbi 1997).

Polyglutamine Expansion Confers Novel in vitro Properties to the Target Proteins

The finding that the abnormal proteins present in patients with HD, SCA1, SCA2, SCA3 and SCA7 are selectively recognized by a monoclonal antibody (1C2) provided additional support for a common pathogenetic mechanism, implicating an abnormal conformation of elongated polyglutamines (Trottier et al. 1995b). This mAb detects huntingtin by Western blot analysis of whole cell protein extracts above a threshold of about 33–38 gln, and its affinity increases with polygln length. Furthermore, truncated proteins containing expanded polygln also show in vitro a tendency to aggregate into insoluble material, forming rather homogeneous, amyloid-like fibrils (Lunkes et al. 1999; Scherzinger et al. 1997). It had been previously proposed that polyglutamine tracts can oligomerize and form β pleated sheets (Stott et al. 1995). The kinetic of aggregation is dependent on polygln length, time and concentration and is accelerated by seeding with preformed fibrils (Scherzinger et al. 1999). It also depends on the nature of the protein; it was shown that a GST fusion with the N-terminal region of mutated huntingtin aggregates only after cleavage of the GST moiety (Scherzinger et al. 1997). The striking similarity of polygln threshold of recognition by 1C2 and of aggregation properties, with the pathological threshold around 36–40 gln in five diseases, suggests that these properties reflect features important for pathogenesis.

Nuclear Inclusions and Mechanisms of Neurodegeneration

Attention was called to nuclear inclusions in polyglutamine diseases, following observations made on a mouse model for HD (R6 lines), where expression of a very short truncated huntingtin fragment carrying 115–155 gln (a length corresponding to a disease onset in the first years of life) was leading to a strong neurological phenotype and significant brain shrinkage (Mangiarini et al. 1996). Davies et al. (1997) reported the conspicuous presence, in some but not all areas of the brain of these mice, of nuclear inclusions (NIs) stained with a anti-huntingtin antibody and also showing ubiquitin immunoreactivity. In some regions (that correspond to those affected in Huntington's disease), it was estimated that most neurons contain in NI. A similar observation of NIs was made in another mouse model, where overexpression in Purkinje cells of a full length ataxin 1 containing 82 gln (also corresponding to an infantile onset) leads to severe ataxia, cerebellar dysfunction and ultimately Purkinje cell death (Burright et al. 1995; Skinner et al. 1997). In both cases, the appearance of the nuclear inclusions preceded by several weeks the onset of clinical symptoms. Finally, Drosophila models have been created by overexpression of long polygln containing constructs in the developing eye, where intensive eye degeneration is correlated with the formation of NIs (Kazemi-Esfarjani and Benzer 2000; Warrick et al. 1998).

Both mouse models have some built-in artificial features: expression of a truncated protein (Mangiarini et al. 1996) or gross overexpression of the full length protein (Burright et al. 1995). However, similar observations were made on patients' brains in HD and five other diseases (SCA1, 2, 3 and 7, DRPLA and SBMA; Becher et al. 1998; Di Figlia et al. 1997; Koyano et al. 2000; Paulson et al. 1997). In HD, NIs were observed in the cortex and in medium-sized neurons of the striatum, but not in non-affected brain areas or in controls. These NIs are stained with an antibody specific for the N-terminal region of huntingtin (very close to the polyglutamine tract) and by an anti-ubiquitin antibody, but not by

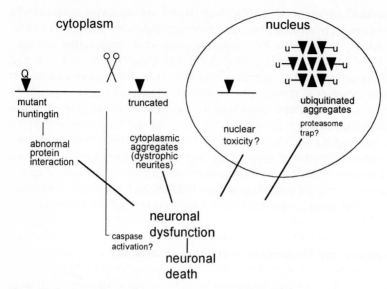

Fig. 1. Mechanisms of neurodegeneration in Huntington's disease

antibodies recognizing other regions of huntingtin (Di Figlia et al. 1997). In one study the NIs appeared much rarer in adult onset cases, and another type of pathological immunoreactive aggregate was observed within dystrophic neurites (distended axon terminals or dendrites; Di Figlia et al. 1997). It was recently proposed that the distribution of dystrophic neurites correlates better with the sites of neurodegeneration than for NIs (Gutekunst et al. 1999). It is surprising at first sight that huntingtin or ataxin-3, which are cytoplasmic proteins, appears to aggregate in the nucleus. At least in the case of huntingtin, only a short N-terminal fragment accumulates in the NIs (Becher et al. 1998; Di Figlia et al. 1997). Thus, one may hypothesize that processing of the mutated protein leads to the production of such a fragment, which would be transported to the nucleus where the environment would favor aggregation (Fig. 1). Subsequent ubiquitination would reflect a failed attempt to degrade the NIs. This is supported by the general observation that several proteasome subunits and chaperones colocalize with NIs. Recent observations in cellular models indicate that transport to the nucleus is polygln dependent, as short truncated fragments with normal polygln length remain localized to the cytoplasm (Lunkes and Mandel 1998). It was generally observed in mouse and cellular models that truncated proteins have higher aggregation potential and toxicity than the full-length proteins.

    If the presence of NIs is undoubtedly correlated to the diseases, their direct contribution to the toxic process is not yet clarified. Three model systems provide examples of a possible dissociation of polyglutamine toxicity from NIs. Transgenic SCA1 mice were generated in which the transgene construct harbored, in addition to the CAG expansion, a small deletion of a region coding for a self association domain of ataxin-1. Purkinje cells overexpressing this protein developed a similar pathological process as the one affecting the original SCA1 mouse model, but did not present NIs (Klement et al. 1998; Lin et al. 2000). Moreover, a study on primary neuronal cells transfected with various types of HD constructs could not correlate the rate of cell death with the proportion of NI-containing cells (Saudou et al. 1998). Finally, in Drosophila models, when the polyglutamine-mediated neurodegeneration was suppressed by the overexpression of protein chaperones from the Hsp40 or 70 families, rescue occurred while NIs were still present (Kazemi-Esfarjani and Benzer 2000; Warrick et al. 1999).

Current observations strongly support the belief that polyglutamines induce cellular dysfunctions before cell death. This situation is difficult to study in man as most samples studied are from patients at the end of many years of disease progression. The mouse models are thus invaluable. In SCA1 mice, the ataxia phenotype and decreased arborization of Purkinje cells occur long before loss of Purkinje cells (Clark et al. 1997). In the R6 HD model, although the neurological symptomatology is very severe, there are only discrete signs of neuronal death and no widespread astrogliosis. The 20 % brain weight loss observed in these mice is currently unexplained (Mangiarini et al. 1996; Turmaine et al. 2000). Moreover, we do not observe massive cell loss in our SCA7 mouse model (see below); however, it displays very severe functional and histological alterations. To analyze the dysfunctional state of affected neurons, expression profiling analysis or use of subtracted cDNA libraries has been performed (Lin et al. 2000; Luthi-Carter et al. 2000). Downregulation of numerous genes was observed at early stages of pathogenesis. This finding suggests that transcriptional alterations may be part of the pathological mechanisms, which may be correlated to interaction between mutated huntingtin and the transcriptional factor CBP (Steffan et al. 2000).

It has also been proposed that activation of some caspases may contribute to neuronal dysfunction (Goldberg et al. 1996; Ona et al. 1999).

The Mystery of the Specificity of Neurodegeneration

The ubiquitous pattern of expression of the target proteins in the various diseases contrasts with the regional specificity of neurodegeneration. For instance, huntingtin is expressed at high levels in Purkinje cells that are not affected in adult onset HD. However, the same cells are the most sensitive to the SCA1 mutation. It has been proposed that abnormal interactions between the mutated protein and some neuron-specific proteins could account for the selective pathology. Using a genetic strategy (the yeast two-hybrid system) to screen for protein interacting with ataxin 1, Matilla et al. (1997) have found a leucine-rich protein (LANP) that interacts much better with an ataxin 1 with 82 repeats than with a normal protein with 30 gln. This protein is expressed predominantly in Purkinje cells (the primary pathological target in SCA1) and colocalizes with ataxin 1 in the nuclear matrix. The authors propose that the presence of mutated ataxin may interfere with the normal function of the LANP protein (Matilla et al. 1997). Another indication that nuclear matrix may be affected is provided by the abnormal distribution of the PML protein (a nuclear matrix protein associated with nuclear bodies) in Cos cells overexpressing mutated ataxin (Skinner et al. 1997). It remains to be seen whether other proteins showing sensitivity to polyglutamine length and appropriate neuronal specificity of expression can be found to interact with mutated proteins in other diseases. Proteins whose interaction with huntingtin are modulated by the number of glutamines have been described (HAP1 HIP1, the transcription factor CBP and others), but they do not show a specificity of expression in the neurons selectively affected in HD (Li et al. 1996; Steffan et al. 2000; Wanker et al. 1997; for a review of proposed interactors of huntingtin, see Jones 2000). It should be noted, however, that NIs and a neurological phenotype (withour obvious neurodegeneration) were obtained by inserting a long CAG/polyglutamine repeat within the mouse HPRT (Hypoxanthin Phosphoribosyl Transferase) gene (Ordway et al. 1997). This gene is not associated with a polyglutamine expansion disease in man. This finding suggests that the existence of specific interactors may not be mandatory to elicit polyglutaminemediated neuronal dysfunction.

Alternate (not necessarily exclusive) explanations for the regional selectivity could lie in subtle differences in expression of the target proteins in different neuronal populations, or in differences in the processing of these proteins, that would produce truncated fragments with a higher aggregation potential or toxicity in some neurons. An argument in favor of the first hypothesis is provided by the similarity in the regions affected in HD patients and in the R6 HD mouse model (that express a highly truncated protein under the control of the promoter region of the human HD gene; Davies et al. 1997). It was also reported that within the striatum, the selective vulnerability of different neuron types appears correlated with the normal level of huntingtin expression in the corresponding rat neurons (Kosinski et al. 1997). It is well known that the kinetics of protein aggregation can be highly influenced by rather small differences in protein concentration (Jarrett and Lansbury 1993; Scherzinger et al. 1999). On the other hand, it has been suggested that the expanded polyglutamine may lead to increased sensitivity of huntingtin to cleavage by apoptosis-related proteases (caspase 3; Goldberg et al. 1996). However, the cleavage site accounting for the truncated fragments that accumulates in NIs is upstream of the caspase 3 site of huntingtin (Lunkes and Mandel 1998; see below).

An Inducible Cellular Model for Analysis of Proteolytic Processing of Huntingtin

We have devised a cellular model where expression of full length or truncated huntingtin, with normal or expanded polygln repeats, is under the control of a tetracycline inducible promoter (through binding of tetracycline or its analogue doxycycline to a bacterial transactivator; Lunkes and Mandel 1998). The cell line we have chosen is a neuroblastoma/glioma hybrid line (NG108) that expresses a neuron-like phenotype upon differentiation. These cells can be studied at least 16 days after induction of huntingtin expression and differentiation, i. e., much longer than in transient transfection systems (for a review on cellular models, see Lunkes and Mandel 2000).

Using this model, we could show the progressive formation of nuclear and cytoplasmic aggregates, depending on the length of the recombinant protein (full-length or truncated) and of the polygln tract. In this system, NIs were efficiently formed after 12 days from full-length huntingtin carrying 116 glutamines, whereas very few NIs but more cytoplasmic inclusions (CIs) were formed with the same construct carrying only 73 repeats. However, the polyglutamine-containing CIs and NIs contain only an N-terminal fragment of huntingtin, detected by antibodies against epitopes flanking on either side the polyglutamine repeat. A monoclonal antibody (4C8) recognizing an epitope proximal and upstream of the caspase-3 cleavage site does not detect the inclusions in these cells, or in patients's brains. This finding indicates that a cleavage occurs at a site proximal to the caspase-3 site, both in the cell model and in patients. We are using this system to further locate the precise cleavage site and identify the responsible protease, as it may provide a potential therapeutic target.

Modelling Polyglutamine Toxicity in the Mouse Retina

Among the polyglutamine disorders, SCA7 is the only one to display degeneration in both brain and retina. Furthermore, it shows the greatest sensitivity to polyglutamine length (apart from SCA6, where a different toxic mechanism may be involved; see above). This appears favorable for generating detectable pathology over the life span of the mouse, using a full-length construct. We have thus generated transgenic mice overexpressing full-length human ataxin-7 (harboring 10 or 90 glutamines) in rod photoreceptors by using the rhodopsin promoter. In this model, expression of the mutant protein causes vision deficiencies and early onset degenerative changes in the retina.

We have raised antibodies against both extremities of human ataxin-7 and showed that an N-terminal immunoreactivity progressively accumulates in the nucleus of photoreceptors, leading to the formation of NIs. These NIs are ubiquitinated and recruit a distinct set of chaperone/proteasome subunits. NIs are not immunoreactive for C-terminal ataxin-7 antibodies, suggesting a proteolytic processing of ataxin-7[Q90]. These molecular features are not attributable to the particularities of the photoreceptor cell type: we could confirm them in another SCA7 mouse model that expresses the protein in Purkinje cells. Such a proteolytic step may therefore be involved in SCA7 as well as in HD and DRPLA (Di Figlia et al. 1997; Lunkes and Mandel 1998; Schilling et al. 1999). Overexpression of ataxin-7[Q90] in rods caused severe retinal degeneration, including loss of outer segments and disorganization of the outer nuclear layer. Pathological changes appeared at about 1 month of age and were correlated to an abnormal electroretinogram response in vivo. Alterations were not limited to the targeted cell type: horizontal and rod-bipolar cells, the post-synaptic partners of rods, developed morphological changes, although they did not express the mutation.

These observations are the first clear evidence that trans-neuronal responses are part of polyglutamine toxicity. Because retina is a suitable tissue for gene transfer and drug delivery, our model offers new opportunities to test therapeutic strategies for triplet repeat disorders.

New Hopes for Therapy

Recent discoveries have raised exciting new hopes for treatment that may delay or suppress the polyglutamine-mediated toxicity. Several neurotrophic factors and anti-apoptotic agents have shown protective effects on cellular models (see Lunkes and Mandel 2000 for a review). The aggregation properties of mutated proteins also provide a potential target, and efforts have been directed towards identification of small molecules that could decrease fibrillary formation in vitro (Heiser et al. 2000). The availability of animal models allows in vivo assessment of therapeutic approaches. One study reported that the neurological phenotype of R6 HD mice can be delayed by crossing them with transgenic animals expressing a dominant negative form of caspase-1, or by treatment with a caspase inhibitor (Ona et al. 1999). Further hopes arise from fly geneticists. Two groups have reported that when flies express either hsp70 or hsp40 chaperones together with a polyglutamine-coding transgene, the severe neurodegenerative phenotype of the model is suppressed (Kazemi-Esfarjani and Benzer 2000; Warrick et al. 1999). However, treatments may encounter enormous difficulties if the disease progression is irreversible after its onset. A recent report based on an inducible mouse model showed that even after the appearance of symptoms, if the expression of the HD mutation is turned off, the animals recover (Yamamoto et al. 2000). This observation suggests that the pathogenic process is not irreversibly triggered by the mutation but is rather continuous and therefore is more likely to be blocked by effective treatments.

Acknowledgments

Studies in our laboratory are supported by CNRS, INSERM, CHRU Strasbourg and Fondation Louis Jeantet. A.L. and G.Y. are supported by the Hereditary Disease Foundation and the Fondation pour la Recherche Medicale, respectively.

References

Becher MW, Kotzuk JA, Sharp AH, Davies SW, Bates GP, Price DL, Ross CA (1998) Intranuclear neuronal inclusions in Huntington's disease and dentatorubral and pallidoluysian atrophy: correlation between the density of inclusions and *IT15* CAG triplet repeat length. Neurobiol Dis 4: 387–397.

Brinkman RR, Mezei MM, Theilmann J, Almqvist E, Hayden MR (1997) The likelihood of being affected with Huntington's disease by a particular age, for a specific CAG size. Am J Human Genet 60: 1202–1210.

Burright EN, Clark HB, Servadio A, Matilla T, Feddersen RM, Yunis WS, Duvick LA, Zoghbi HY, Orr HT (1995) SCA1 transgenic mice: a model for neurodegeneration caused by an expanded CAG trinucleotide repeat. Cell 82: 937–948.

Clark HB, Burright EN, Yunis WS, Larson S, Wilcox C, Hartman B, Matilla A, Zoghbi HY, Orr HT (1997) Purkinje cell expression of a mutant allele of SCA1 in transgenic mice leads to disparate effects on motor behaviors followed by a progressive cerebellar dysfunction and histological alterations. J Neurosci 17: 7385–7395.

David G, Abbas N, Stevanin G, Durr A, Yvert G, Cancel G, Weber C, Imbert G, Saudou F, Antoniou E, Drabkin H, Gemmill R, Giunti P, Benomar A, Wood N, Ruberg M, Agid Y, Mandel JL, Brice A (1997) Cloning of the SCA7 gene reveals a highly unstable CAG repeat expansion. Nature Genet 17: 65–70.

Davies SW, Turmaine M, Cozens BA, Difiglia M, Sharp AH, Ross CA, Scherzinger E, Wanker EE, Mangiarini L, Bates GP (1997) Formation of neuronal intranuclear inclusions underlies the neurological dysfunction in mice transgenic for the HD mutation. Cell 90: 537–548.

Di Figlia M, Sapp E, Chase KO, Davies SW, Bates GP, Vonsattel JP, Aronin N (1997) Aggregation of Huntington in neuronal intranuclear inclusions and dystrophic neurites in brain. Science 277: 1990–1993.

Goldberg YP, Nicholson DW, Rasper DM, Kalchman MA, Koide HB, Graham RK, Bromm M, Kazemi-Esfarjani P, Thornberry NA, Vaillancourt JP, Hayden MR (1996) Cleavage of huntingtin by apopain, a proapoptotic cysteine protease, is modulated by the polyglutamine tract. Nature Genet 13: 442–449.

Gusella JF, McNeil S, Persichetti F, Srinidhi J, Novelletto A, Bird E, Faber P, Vonsattel JP, Myers RH, MacDonald ME (1996) Huntington's disease: Cold Spring Harbor symposia on quantitative biology.

Gutekunst CA, Li SH, Yi H, Mulroy JS, Kuemmerle S, Jones R, Rye D, Ferrante RJ, Hersch SM, Li XJ (1999) Nuclear and neuropil aggregates in Huntington's disease: relationship to neuropathology. J Neurosci 19: 2522–2534.

Harper PS (1991) Huntington's disease. London, W. B. Saunders.

Harper PS (1992) The epidemiology of Huntington's disease. Human Genet 89: 365–376.

Heiser V, Scherzinger E, Boeddrich A, Nordhoff E, Lurz R, Schugardt N, Lehrach H, Wanker EE (2000) Inhibition of huntingtin fibrillogenesis by specific antibodies and small molecules: implications for Huntington's disease therapy. Proc Natl Acad Sci USA 97: 6739–6744.

Jarrett JT, Lansbury PTJ (1993) Seeding "one-dimensional cristallization" of amyloid: a pathogenic mechanism in Alzheimer's disease and scrapie? Cell 73: 1055–1058.

Jones L (2000) Huntingtin-interacting proteins and their relevance to Huntington's disease etiology. Neurosci News 3: 55–63.

Kazemi-Esfarjani P, Benzer S (2000) Genetic suppression of polyglutamine toxicity in Drosophila. Science 287: 1837–1840.

Klement IA, Skinner PJ, Kaytor MD, Yi H, Hersch SM, Clark HB, Zoghbi HY, Orr HT (1998) Ataxin-1 nuclear localization and aggregation: role in polyglutamine-induced disease in SCA1 transgenic mice. Cell 95: 41–53.

Koide R, Kobayashi S, Shimohata T, Ikeuchi T, Maruyama M, Saito M, Yamada M, Takahashi H, Tsuji S (1999) A neurological disease caused by an expanded CAG trinucleotide repeat in the TATA-binding protein gene: a new polyglutamine disease? Human Mol Genet 8: 2047–2054.

Kosinski CM, Cha JH, Young AB, Persichetti F, MacDonald M, Gusella JF, Penney JB Jr, Standaert DG (1997) Huntingtin immunoreactivity in the rat neostriatum: differential accumulation in projection and interneurons. Exp Neurol 144: 239–247.

Koyano S, Uchihara T, Fujigasaki H, Nakamura A, Yagishita S, Iwabuchi K (2000) Neuronal intranuclear inclusions in spinocerebellar ataxia type 2. Ann Neurol 47: 550.

Li XJ, Sharp AH, Li SH, Dawson TM, Snyder SH, Ross CA (1996) Huntingtin associated protein (HAP1): Discrete neuronal localizations in the brain resemble those of neuronal nitric oxide synthase. Proc Natl Acad Sci USA 93: 4839–4844.

Lin X, Antalffy B, Kang D, Orr HT, Zoghbi HY (2000) Polyglutamine expansion down-regulates specific neuronal genes before pathologic changes in SCA1. Nature Neurosci 3: 157–163.

Lunkes A, Mandel J-L (1998) A cellular model that recapitulates major pathogenic steps of Huntington's disease. Human Mol Genet 7: 1355–1361.

Lunkes A, Mandel J-L (2000) Cellular models of Huntington's disease. Neurosci News 3: 30–37.

Lunkes A, Trottier Y, Fagart J, Schultz P, Zeder-Lutz G, Moras D, Mandel J-L (1999) Properties of poly-glutamine expansion in vitro and in a cellular model for Huntington's disease. Phil Trans Roy Soc London [Biol] 354: 1013–1019.

Luthi-Carter R, Strand A, Peters NL, Solano SM, Hollingsworth ZR, Menon AS, Frey AS, Spektor BS, Penney EB, Schilling G, Ross CA, Borchelt DR, Tapscott SJ, Young AB, Cha JH, Olson JM (2000) Decreased expression of striatal signaling genes in a mouse model of Huntington's disease. Human Mol Genet 9: 1259–1271.

Mangiarini L, Sathasivam K, Seller M, Cozens B, Harper A, Hetherington C, Lawton M, Trottier Y, Leh-rach H, Davies SW, Bates GP (1996) Exon 1 of the HD gene with an expanded CAG repeat is suffi-cient to cause a progressive neurological phenotype in transgenic mice. Cell 87: 493–506.

Matilla A, Koshy B, Cummings CJ, Isobe T, Orr HT, Zoghbi HY (1997) The cerebellar leucine rich acidic nuclear protein (LANP) interacts with ataxin-1. Nature 389: 974–978.

Ona VO, Li M, Vonsattel JP, Andrews LJ, Khan SQ, Chung WM, Frey AS, Menon AS, Li XJ, Stieg PE, Yuan J, Penney JB, Young AB, Cha JH, Friedlander RM (1999) Inhibition of caspase-1 slows disease progression in a mouse model of Huntington's disease. Nature 399: 263–267.

Ordway JM, Tallaksen-Greene S, Gutekunst CA, Bernstein EM, Cearley JA, Wiener HW, Dure LS, Lind-sey R, Hersch SM, Jope RS, Albin RL, Detloff PJ (1997) Ectopically expressed CAG repeats cause intranuclear inclusions and a progressive late onset neurological phenotype in the mouse. Cell 91: 753–763.

Paulson HL, Perez MK, Trottier Y, Trojanowski JQ, Subramony SH, Das SS, Vig P, Mandel JL, Fisch-beck KH, Pittman RN (1997) Intranuclear inclusions of expanded polyglutamine protein in spino-cerebellar ataxia type 3. Neuron 19: 333–344.

Rubinsztein DC, Leggo J, Chiano M, Dodge A, Norbury G, Roser E, Craufurd D (1997) Genotypes at the GluR6 kainate receptor locus are associated with variation in the age of onset of Huntington's dis-ease. Proc Natl Acad Sci USA 94: 3872–3876.

Saudou F, Finkbeiner S, Devys D, Greenberg ME (1998) Huntingtin acts in the nucleus to induce apop-tosis but death does not correlate with the formation of intranuclear inclusions. Cell 95: 55–66.

Scherzinger E, Lurz R, Turmaine M, Mangiarini L, Hollenbach B, Hasenbank R, Bates GP, Davies SW, Lehrach H, Wanker EE (1997) Huntingtin-encoded polyglutamine expansions form amyloid-like protein aggregates in vitro and in vivo. Cell 90: 549–558.

Scherzinger E, Sittler A, Schweiger K, Heiser V, Lurz R, Hasenbank R, Bates GP, Lehrach H, Wanker EE (1999) Self-assembly of polyglutamine-containing huntingtin fragments into amyloid-like fibrils: implications for Huntington's disease pathology. Proc Natl Acad Sci USA 96: 4604–4609.

Schilling G, Wood JD, Duan K, Slunt HH, Gonzales V, Yamada M, Cooper JK, Margolis RL, Jenkins NA, Copeland NG, Takahashi H, Tsuji S, Price DL, Borchelt DR, Ross CA (1999) Nuclear accumulation of truncated atrophin-1 fragments in a transgenic mouse model of DRPLA. Neuron 24: 275–286.

Skinner PJ, Koshy B, Cummings CJ, Klement IA, Helin K, Servadio A, Zoghbi HY, Orr HT (1997) SCA1 pathogenesis involves alterations in nuclear matrix associated structures. Nature 389: 971–974.

Steffan JS, Kazantsev A, Spasic-Boskovic O, Greenwald M, Zhu YZ, Gohler H, Wanker EE, Bates GP, Housman DE, Thompson LM (2000) The Huntington's disease protein interacts with p53 and CREB-binding protein and represses transcription. Proc Natl Acad Sci USA 97: 6763–6768.

Stott K, Blackburn JM, Butler PJ, Perutz M (1995) Incorporation of glutamine repeats makes protein oligomerize: implications for neurodegenerative diseases. Proc Natl Acad Sci USA 92: 6509–6513.

The Huntington's Disease Collaborative Research Group (1993) A novel gene containing a trinucleotide repeat that is expanded and unstable on Huntington's disease chromosomes. Cell 72: 971–983.

Trottier Y, Devys D, Imbert G, Saudou F, An I, Lutz Y, Weber C, Agid Y, Hirsch EC, Mandel JL (1995a) Cellular localization of the Huntington's disease protein and discrimination of the normal and mutated form. Nature Genet 10: 104–110.

Trottier Y, Lutz Y, Stevanin G, Imbert G, Devys D, Cancel G, Saudou F, Weber C, David G, Tora L, Agid Y, Brice A, Mandel J-L (1995b) Polyglutamine expansion as a pathological epitope in Hunting-ton's disease and four dominant cerebellar ataxias. Nature 378: 403–406.

Turmaine M, Raza A, Mahal A, Mangiarini L, Bates GP, Davies SW (2000) Nonapoptotic neurodegener-ation in a transgenic mouse model of Huntington's disease. Proc Natl Acad Sci USA 97: 8093–8097.

Wanker EE, Rovira C, Scherzinger E, Hasenbank R, Walter S, Tait D, Colicelli J, Lehrach H (1997) HIP-I: a huntingtin interacting protein isolated by the yeast two-hybrid system. Human Mol Genet 6: 487–495.

Warrick JM, Paulson HL, Gray-Board GL, Bui QT, Fischbeck KH, Pittman RN, Bonini NM (1998) Expanded polyglutamine protein forms nuclear inclusions and causes neural degeneration in Drosophila. Cell 93: 939–949.

Warrick JM, Chan HYE, Gray-Board GL, Chai Y, Paulson HL, Bonini NM (1999) Suppression of polyglutamine-mediated neurodegeneration in Drosophila by the molecular chaperone Hsp70. Nature Genet 34: 425–428.

White JK, Auerbach W, Duyao MP, Vonsattel J-P, Gusella JF, Joyner AL, MacDonald ME (1997) Huntingtin is required for neurogenesis and is not impaired by the Huntington's disease CAG expansion. Nature Genet 17: 404–410.

Yamamoto A, Lucas JJ, Hen R (2000) Reversal of neuropathology and motor dysfunction in a conditional model of Huntington's disease [In Process Citation]. Cell 101: 57–66.

Zoghbi HY (1997) CAG repeats in SCA6: anticipating new clues. Neurology 49: 1–5.

Prion Protein Biogenesis:
Implications for Neurodegeneration

V. R. Lingappa and R. S. Hegde

Summary

Much attention has been focused on the unusual mode of propagation of the prion protein (PrP). However this is not the only pathophysiologically relevant aspect of PrP biology. Our studies suggest that the mechanism of PrP biogenesis plays a crucial role in the pathway leading to neurodegeneration in prion diseases. First, we have found that, unlike conventional secretory or integral membrane proteins, PrP is synthesized at the endoplasmic reticulum (ER) in multiple topological forms. Second, using transgenic (Tg) mice expressing various PrP mutants on a null background, we have demonstrated that PrP expression in one of these topological forms (termed [Ctm]PrP) is associated with development of spongiform neurodegeneration and astrogliosis. Furthermore, [Ctm]PrP is found in the brains of patients with Gerstmann-Straussler Syndrome, a genetic prion disease. While [Ctm]PrP is not itself infectious, it is induced late in the course of infectious prion disease, suggesting a role for [Ctm]PrP in both genetic and infectious prion disorders. Our studies lead us to the hypothesis that [Ctm]PrP is a transducer of signals for apoptotic neuronal cell death. We also believe that [Ctm]PrP is not simply a pathological molecule but rather may have a physiological role, under certain circumstances. Finally, we suggest that PrP is just one of a family of complex proteins that are regulated at the level of co-translational translocation, whose dysregulation results in disease.

Cell-Free Systems for Protein Biogenesis

Cell-free systems for the study of protein biogenesis at the endoplasmic reticulum (ER) have been in existence for nearly half a century (Palade 1975). In essence, they consist of relatively dilute cytosolic extracts to which are added various components including ATP, GTP, amino acids, transcripts of cloned cDNAs, and subcellular membrane fractions (see Fig. 1). The incorporation of radiolabelled amino acids during protein synthesis in such systems allows the newly synthesized protein, encoded in the cDNA transcript, to be distinguished from all pre-existing proteins by polyacrylamide gel electrophoresis in sodium dodecyl sulphate followed by autoradiography. Although the membrane fractions added are often heterogeneous, the nascent chains will target only to the subset of vesi-

Research and Perspectives in Alzheimer's Diseases
Beyreuther/Christen/Masters (Eds.)
Neurodegenerative Disorders
© Springer-Verlag Berlin Heidelberg 2001

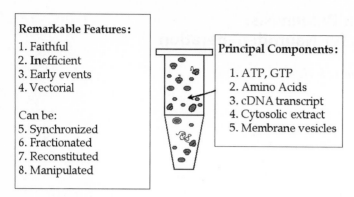

Remarkable Features:
1. Faithful
2. Inefficient
3. Early events
4. Vectorial

Can be:
5. Synchronized
6. Fractionated
7. Reconstituted
8. Manipulated

Principal Components:
1. ATP, GTP
2. Amino Acids
3. cDNA transcript
4. Cytosolic extract
5. Membrane vesicles

Fig. 1. Cell-free systems for protein translocation. To the right are indicated the components of a cell-free translation/translocation system. To the left are indicated some of its key features

cles that are derived from the ER and are in the correct orientation (cytosolic face out). Only these vesicles display the appropriate targeting and translocation receptors in a manner accessible to the radiolabelled nascent chains.

The subsequent addition of proteases such as proteinase K (PK) will digest polypeptides that have not been translocated to the vesicle lumen. When non-denaturing detergents are added during PK treatment, proteins and protein domains that were protected from protease digestion solely by virtue of the lipid bilayer (i.e., on no translocation) are rendered accessible and therefore are degraded (see Blobel and Dobberstein 1975a,b). In the case of integral membrane proteins, the initial transmembrane topology can be determined by combining the proteolytic assay with site-specific antibodies (e.g., raised to specific pep-tides; see for example Schmidt-Rose and Jentsch 1997). Finally, in some cases, compartment-specific post-translational modifications (e.g., N-linked glycosyla-tion, in the case of the ER) can be used to independently corroborate whether a given domain of a newly synthesized protein has crossed a particular membrane (Rothman et al. 1989).

These cell-free systems have two remarkable features (Simon et al. 1987). First, for the most part, they appear to be qualitatively faithful reproductions of events occuring during protein biogenesis in vivo. Second, they are remarkably inefficient, perhaps in part as a result of their relative dilution compared to pro-tein concentrations in vivo. This inefficiency can be greater for some steps of bio-genesis than for others, and thus the distribution of products along the pathway of biogenesis in cell-free systems at equilibrium is typically quantitatively dis-torted, compared to the corresponding events in living cells at steady state.

These features of the cell-free protein synthesizing systems have proven to be an unexpected boon in the analysis of molecular mechanisms. They have allowed protein-protein interactions that are normally too transient or too minor to be detected readily in vivo to be identified, dissected and explored in the cell-free system. In at least one case, that of assembly of viral capsids, cell-free translation systems have allowed detection and study in vitro of transient intermediates.

These findings were corroborated in vivo upon closer scrutiny (Lingappa et al. 1997). Even to the extent that the events in the cell-free system are a quantitative distortion, they allow at least some steps that would be difficult to examine in vivo to be magnified and therefore made more amenable to study.

Other notable features and advantages are summarized in Figure 1. For example, whereas cell fractionation often generates a complex and heterogeneous distribution of a given protein among various compartments, cell-free translocation studies typically give a much simpler outcome. Perhaps this reflects the fact that translocation across the ER membrane is a largely vectorial process, or that only vesicles derived from the ER are competent for de novo translocation, or because transport of proteins significantly beyond the site of synthesis apparently does not occur in these systems under standard conditions. Regardless of the explanation, this feature provides a significant practical distinction between studying a process by de novo biosynthesis in a cell-free system versus by analysis of cells pulse radiolabelled in vivo and subjected to fractionation subsequently. Often it is best to combine both approaches, using cell-free systems to define forms of a protein and then pulse-chase analysis in cells to follow the modifications that occur to those forms in transit through various intracellular compartments.

Finally, the components that are combined to make a functioning cell-free translocation system are amenable to further fractionation and reconstitution. Thus, variations of the cell-free translocation systems have been developed in which most of the proteins in either the ER lumen (Nicchitta and Blobel 1993), the ER membrane (Gorlich and Rapoport 1993) or the cytosol (Hegde et al. unpublished) have been removed. This development allows exploration of the role of specific components in the translocation process and has set the stage for future attempts at reconstitution of regulatory events.

As applied to simple secretory and integral membrane proteins, the cell-free translocation system has contributed decisively to the conventional view of protein biogenesis at the ER membrane. Thus conventional secretory proteins are found in these systems to be fully translocated to the lumenal space, in a reaction that is dependent on the microsomal membranes concentration and allowed testing of the signal hypothesis (Blobel and Dobberstein 1975a, b; Simon and Blobel 1991) and discovery and dissection of the core events of protein targeting (Walter and Johnson 1994) and translocation (Gorlich and Rapoport 1993). Similarly, simple integral membrane proteins are correctly and homogeneously assembled into the bilayer at the ER (Katz and Lodish 1979; Katz et al. 1977).

The data on complex integral membrane proteins, which span the lipid bilayer many times, have been more difficult to assess. The topology achieved in the cell-free system by these newly synthesized proteins has been more heterogeneous than that observed for the simple membrane proteins and has been in some cases, at least partly, at variance with the observations in vivo (Skach et al. 1993). On the one hand, since these proteins have multiple signal and stop transfer sequences, heterogeneity of topology could simply reflect inefficiency of signal or stop transfer sequence engagement early during chain growth, and thus reflect the quantitative distortion discussed above. Alternatively, heterogeneity of multispanning membrane protein topology could reflect a true structural diversity at the level of individual monomers. If, for example, mechanisms exist in vivo (but are not reconstituted in vitro) to switch on or off specific translocation programs, the observed heterogeneity could reflect the range of possibilities from which cells normally select a particular course of action in a programmed manner, for example, under different metabolic or other physiological (or pathological) conditions.

Yet another view is that both cells and cell-free systems are highly ineffecient and make a lot of mistakes. This argument has been used, for example, to account for the fact that the vast majority of wild type Cystic Fibrosis Transmembrane Regulator is degraded immediately upon synthesis, with only a few percent of chains allowed to mature and exit from the ER (Ward and Kopito 1994).

As an approach to understanding the basis for such complexity, as observed both in vitro and in vivo, we have devoted considerable effort over the past two decades to the study of specialized and unusual proteins whose characteristics as translocation substrates in cell-free systems have shed some light on this issue. One of the proteins studied has been the prion protein (PrP). The insights from exploration of PrP biogenesis are summarized here.

Prion Protein Biogenesis

Upon cell-free translation of transcript from a homogeneous full-length PrP cDNA engineered behind the bacteriophage SP6 promoter, a single PrP translation product is observed (see Fig. 2, lane 1). When functional microsomal membranes are present during translation, the nascent PrP chains target to vesicles derived from the ER and initiate translocation. After completion of translation, the topology of the newly synthesized chains can be assessed by digestion with exogenously added proteases in the absence or presence of detergent. When newly synthesized (radiolabeled) PrP is assessed in this fashion, three populations of chains of radically different topology are observed. Some chains, as expected, are fully translocated to the lumenal space, with cleavage of the N-terminal signal sequence (see Fig. 2, lane 3). These chains are termed ^{Sec}PrP because, by the PK digestion assay, their behavior is indistinguishable from that of a secretory protein. In addition, two transmembrane forms of opposite topology are also observed (Hegde et al. 1988a). One, termed ^{Ctm}PrP, has its C-terminus in the ER lumen. The other, termed ^{Ntm}PrP, has its N-terminal domain in the ER lumen and its C-terminal half in the equivalent of the cytoplasm (see Fig. 2).

While the three different forms of PrP were initially characterized by their differences in topology with respect to the ER membrane, a line of experimentation suggests that another way of viewing the differences between the various forms of PrP is that they are different in conformation. The basis for this view is that in non-denaturing detergent solutions (see Fig. 2, lane 4), in which all topological differences are abolished, under conditions in which protein folding should be unperturbed, digestion with PK reveal a distinctly different protease sensitivity on the part of ^{Ctm}PrP compared to either ^{Sec}PrP or ^{Ntm}PrP.

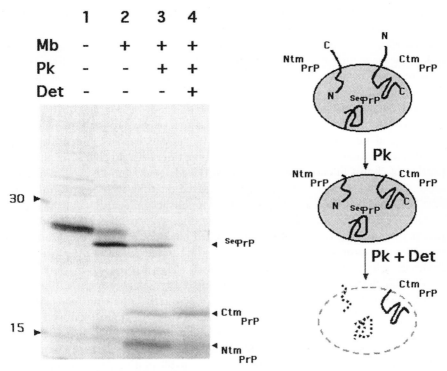

Fig. 2. Analysis of biogenesis of PrP by study in cell-free systems. Details as described in text

Historically, the protease digestion data were initially interpreted to reveal a population of transmembrane chains that spanned the membrane twice, with both N- and C termini in the ER lumen (Hay et al. 1997). However, when monoclonal antibodies became available to a defined epitope just prior to the putative first transmembrane region, it became clear that the N and C terminal protease digestion fragments could not be generated from a single population of chains, since both contained the unique 3F4 epitope, which is only present in one copy per chain.

Several oddities distinctive of PrP biogenesis are worth mentioning, because they further confound the interpretation of protease-digestion patterns, and therefore are a good example of the unexpected way in which quirky features of biological systems can be misleading when interpreted by the simplest hypothesis.

First, even though the 3F4 epitope is on the cytosolic side in the case of CTMPrP chains, it is protected from protease digestion, presumably due to the polypeptide's distinctive conformation (see below). This position is fortunate, because otherwise it would not have been possible to conclude that the N and C terminal fragments (both of which contain the 3F4 epitope) were derived from different populations of chains.

Secondly, for reasons currently not understood, at least in cell-free systems, the signal sequence appears not to be cleaved from PrP chains in the Ctm configuration. One possibility is that this reflects the regulatory state of the ER, with lack of cleavage due to a decision by the cell to proceed with degradation rather than maturation of [Ctm]PrP. Of course, a more trivial possibility is that lack of cleavage of the [Ctm]PrP signal sequence is simply a reflection of the inefficiency of the cell-free system. In either case, we know that lack of cleavage of the signal sequence is not due to lack of access of the signal peptide to the lumenally disposed signal peptidase, since a cleavage site mutant has been identified that results in quantitative cleavage of the signal peptide, with no change in the proportion of chains in each topological form (Hegde et al., unpublished). Furthermore, the signal sequence appears to be cleaved quantitatively from CtmPrP in vivo (Hegde et al. 1998a and Hegde et al., unpublished). Regardless of the explanation, the fact that the signal sequence is uncleaved in the cell-free systems results in CtmPrP chains synthesized in vitro having a distincitve migration in the absence of protease, when the chains are glycosylated, and in the presence of protease, when chains are not glycosylated. Thus, when glycosylated, the CtmPrPPK cleavage product migrates anomalously, roughly comigrating with signal-cleaved full-length secretory chains. Conversely, when not glycosylated, full-length CtmPrP comigrates with signal-uncleaved PrP chains that have failed to target to the ER.

Since PrP is known to have a glycolipid anchor (Stahl et al. 1987), and since N-terminal signal sequences are known to be capable of directing glycolipidation (Caras and Weddell 1989), it is possible that the CtmPrP chains are directed by engagement of the ER membrane by the C terminil of PrP chains whose N termini fail to target correctly. However, two lines of experimentation make this scenario unlikely. First, such a model would predict that the distribution of PrP topologic forms would change with membrane concentration. As membrane concentration increased, the fraction of chains targeting via the N-terminal signal sequence would increase, and the percentage of Ctm chains would decrease. This result process is not observed (Rutkowski et al., unpublished). Furthermore, by that hypothesis, termination of PrP chains at the serine to which glycolipidation would normally occur would be predicted to alter the distribution of topological isoforms. This result is not observed either, suggesting that the glycolipid anchor sequence is not directly involved in topological regulation of PrP (Lingappa et al., unpublished).

The initial description of transmembrane form(s) of PrP was met with considerable skepticism because the relevance of these observations to the in vivo situation had not been established. Of course, analysis of the distribution of topological forms in scrapie-infected brains was problematical because of the protease restistance of PrP[Sc] made protease-based scoring of topology impossible. Thus, the "noise" of harsh proteinase K-resistant PrP[Sc] was too great to allow detection of small amounts of [Ctm]PrP.

Transmembrane PrP in Genetic Prion Disease

An important breakthrough was the identification of mutations that skewed the distribution of PrP topology towards or away from ^{Sec}PrP, the form normally observed in vivo. When mutations that favored ^{Ctm}PrP were introduced as transgenes into mice whose endogenous PrP genes had been deleted by homologous recombination, a dramatic neurodegenerative syndrome was observed (Hegde et al. 1998a). Characterized by intense, focal astrogliosis and spongiform change, this disease had the pathological hallmarks of prion disease. However, analysis of brain tissue revealed no PrP^{Sc} and the brain homogenates were subsequently found to have substantial amounts of ^{Ctm}PrP but no detectable PrP^{Sc}. Furthermore, the brain homogenates from these sick mice were found to be completely non-infectious. Thus, either ^{Ctm}PrP-containing disorders represented a completely different form of prion disease or the steps involving ^{Ctm}PrP are distinct from those of PrP^{Sc}. Finally, the demonstration that ^{Ctm}PrP in these mouse brains was almost exclusively endo H resistant suggested strongly that these molecular had passed "quality control" and had been exported out of the ER, presumably to a distal compartment. Thus, ^{Ctm}PrP-mediated neurodegeneration was not an abberation of protein mislocalization in the most trivial sense. More likely, signaling mechanisms to monitor ^{Ctm}PrP levels instruct the quality control machinery to allow the exit of ^{Ctm}PrP when it reaches a certain concentration.

A Final Common Pathway of Neurodegeneration in Both Genetic and Infectious Prion Diseases

To explore the relationship of PrP^{Sc} to ^{Ctm}PrP, two lines of experimentation were initiated (Hedge et al. 1999). First, we compared various mutations that favored CTMPrP upon cell-free translation versus upon expression in transgenic (Tg) mice. We found a striking correlation between the propensity of a mutation to favor CTMPrP in the cell-free system, adjusted for the total level of transgene expression, and the time to spontaneous clinical illness in Tg mice. From this a so-called Ctm-index was derived that related the propensity of a mutation to favor synthesis as ^{Ctm}PrP, adjusted for the level of transgene expression. Then, the various lines of Tg mice were inoculated with scrapie, and the level of PrP^{Sc} and the time to clinical disease were quantified. An inverse relationship was observed between the level of PrP^{Sc} observed at the time of illness and the Ctm-index. Thus, the animals that generated the highest levels of PrP^{Sc} were the ones that came down with clinical disease the most slowly and had the least propensity to form ^{Ctm}PrP. This finding suggested that ^{Ctm}PrP was a transducer of the signal from PrP^{Sc} that leads to neurodegeneration but was not itself either a precursor or product of PrP^{Sc}.

A second experimental approach was to study PrP null mice carrying both the wild-type (wt) Syrian hamster (Sha) and wt mouse PrP transgenes (generated by crossing PrP null mice bearing the individual wt mouse and Sha PrP transgenes). Here we took advantage of the fact that there is no apparent species

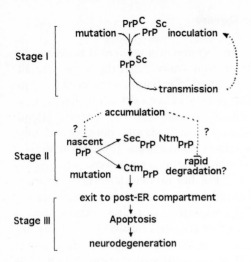

Fig. 3. Model of the relationship of PrP biogenesis to neurodegeneration. The role of PrP^{Sc} is to convert normal molecules of PrP (presumably ^{Sec}PrP) into "infectious" molecules of PrP^{Sc}. As PrP^{Sc} accumulates, it signals the ER to either turn off TrAF and/or turn off ER degradation, resulting in exit of ^{Ctm}PrP from the ER. Upon reaching its mature subcellular location, ^{Ctm}PrP appears to trigger neurodegeneration via an apoptotic mechanism

barrier in generation of ^{Ctm}PrP (i.e., both mouse and Sha wt PrP have an equal propensity to generate transmembrane chains upon cell-free translation studies), in contrast to the well-described species barrier to production of PrP^{Sc} (DeArmond and Prusiner 1995). Thus, if the double Tg mice are inoculated with mouse prions, mouse PrP will be converted to PrP^{Sc}. However, the species barrier will prevent hamster PrP from being converted into PrP^{Sc}. Under these conditions it becomes possible to test the hypothesis of a signaling pathway by which PrP^{Sc} triggers neurodegeneration via ^{Ctm}PrP. If true, inoculation of the double Tg animals with mouse prions should direct the synthesis of ^{Ctm}PrP. The lack of hamster PrP^{Sc} and the existence of hamster-specific monoclonal antibodies would allow the hamster transgene to serve as a reporter of signals for newly synthesized PrP to be made in the form of ^{Ctm}PrP made from the hamster transgene (^{Ctm}PrP made from the mouse transgene would be obscured by the presence of protease-resistant mouse PrP^{Sc}). Those experiments revealed that ^{Ctm}PrP was indeed observed during the course of scrapie infection, rising above background just prior to the onset of clinical symptoms. Thus the data also support the more general hypothesis that ^{Ctm}PrP represents a crucial step in the pathogenesis of both genetic and infectious prion diseases (See Fig. 3).

Molecular Mechanisms by Which PrP^{Sc} Might Trigger ^{Ctm}PrP Expression

The core translocation machinery by which the simplest secretory and integral membrane proteins are transported to the ER lumen is known (Gorlich and Rapoport 1993). It is composed of the trimeric Sec 61 complex and the two-subunit Signal Recognition Particle Receptor (SRPR). Another polypeptide, termed the Translocon-associated Membrane protein (TRAM) is needed for translocation of some secretory proteins.

When PrP is translated in the presence of reconstituted membranes containing the minimum components, with or without the addition of TRAM, exclusively ^{Ctm}PrP is synthesized (Hegde et al. 1998b). Futhermore, mutants that make exclusively ^{Sec}PrP are unable to be translocated across minimal membranes, despite the fact that these membranes are fully competent for translocation of simple secretory and intregral membrane proteins. When a glycoprotein fraction is added back during membrane reconstitution, the translocon becomes competent to synthesize ^{Sec}PrP. However, reconstitution with purified TRAM, a glycoprotein, does not confer competence to synthesize ^{Sec}PrP. Thus, glycoproteins other than TRAM are necessary and comprise a Translocation accessory factor (TrAF) needed for ^{Sec}PrP synthesis. Since ^{Ctm}PrP is the default form made in its absence, TrAF can be viewed as a protective factor whose absence or inactivation could trigger neurodegeneration due to enhanced synthesis of ^{Ctm}PrP.

Regulation at the level of TrAF is not the only mechanism by which ^{Ctm}PrP could be activated to initiate a neurodegenerative cascade. It is well known that the ER has a robust quality control machinery that binds many proteins (Ellgaard et al. 1999), preventing their movement out of the ER and in some cases returning them to the cytosol for rapid degradation (Wiertz et al. 1996). Some evidence suggests that signaling from the cell surface is capable of turning off quality control machinery, including degradation, providing a means by which a protein or protein complex could leave the ER at some times but not at others (Bonifacino et al. 1990). By analogy to studies on T Cell Receptor degradation versus assembly and exit from the ER, it is possible that regulation of ^{Ctm}PrP could occur, all or in part, at the level of degradation versus export, rather than exclusively at the level of synthesis. By analogy to other complex signaling systems, it is likely that events at the level of synthesis reinforce events at the level of degradation versus exit. Thus, those conditions under which TrAF is physiologically inactivated might be expected to also inactivate degradation in order to allow ^{Ctm}PrP to exit the ER efficiently. On the other hand it is possible that, under certain conditions, in response to either defective TrAF or degradation activity, the other level is upregulated as a homeostatic response. Further studies will be necessary to determine if TrAF and ER degradation are regulated in tandem as is the case for other dimensions of ER quality control.

Implications and Future Directions

Our studies in cell-free systems and in Tg mice, described above, have revealed a previously unappreciated dimension of translocational regulation at the ER that is highly relevant to both genetic and infectious prion desease pathogenesis. However, a number of fundamental unanswered questions remain to be explored, some of which are briefly indicated here.

First, and most practically, the currently unknown mechanism by which PrP biogenesis is related to PrP function is likely to be more complex than presented in the simplified current models (see Fig. 3). These models are useful to indicate new conceptual frameworks but should not be taken too literally. We do not yet know if an [Ntm]PrP form exists in brain, and if so, its relationship to the other topological forms observed. Additional bands are observed upon proteolysis of brain fractions and homogenates. Are these still other toplogical or conformational forms, or intermediates in maturation through the secretory pathway? How do they relate to the much simpler pattern observed in cell-free systems?

Second, viewed more broadly, we do not understand whether translocational regulation is a general property of most complex secretory and membrane proteins or a specialized feature limited to very few. Several suggestive, but not conclusive, lines of evidence support the former, more general, hypothesis. Our studies of other complex proteins completely unrelated to PrP, including apolipoprotein B, MDR-1 p-glycoprotein, and other polypeptides, suggest that translocational regulation is a general property of higher eukaryotes, although it is manifest in very different ways, from protein to protein (Skach et al. 1993; Hegde and Lingappa 1996; and unpublished data). The microsomal membranes of dog pancreas in which TrAF is found are not normally engaged in significant PrP synthesis, suggesting that TrAF exists to regulate biogenesis of other, as yet unknown, pancreatic proteins. Finally, studies using microsomal membranes from more primitive eukaryotes (e.g., species that lack a nervous system) have demonstrated that the ER is devoid of TrAF activity, suggesting that TrAFs and translocational regulation may be a relatively recently evolved function, perhaps related to and necessary for the evolution of complex organ systems such as the brain and immune system (Lingappa and Lucero, unpublished).

The molecular mechanism by which nascent chains are allocated to different pathways of biogenesis is unknown. Why do [Sec]PrP-favoring mutants fail to translocate across the ER membrane? Why is [Ctm]PrP the default in the absence of TrAF? In the case of PrP the phenotype by which these mechanisms are manifest is topological. For other proteins, phenomena that are essentially analogous at the molecular level may manifest phenotypically as a change in protein conformation. Recent studies suggest that such conformational regulation may be a general and largely unappreciated feature of the function of the ER (Rutkowski et al., unpublished).

Regardless of the molecular mechanism of translocational regulation, how does it connect to other regulatory mechanisms of the ER, including the functioning of molecular chaperones (Klappa et al. 1998), the unfolded protein response pathway (Sidrauski et al. 1998), and the pathway(s) of ER degradation, including those mediated by the proteasome (Wiertz et al. 1996).

In addition to these cell biological questions, in which the prion is viewed as a specific case of a more general biological regulatory mechanism, there are important, neurodegeneration-specific issues. Perhaps foremost among these is whether [Ctm]PrP is a final common pathway not only for prion-induced neurodegeneration, but also for other forms of neurodegeneration. Thus, everything from stroke to Parkinson's Disease or Alzheimer's Disease might, like prion disease, be an example of signaling in which the actual cause of neurodegeneration is exit of ^{Ctm}PrP from the ER. These possibilities can be explored for any disease for which a mouse model existes by experiments analogous to the double Tg mouse experiments described above. Such a connection would be supported if the disease time course is altered (either shortened or lengthened) in either PrP null mice or in mice favoring one topological form of PrP over any other. Such experiments are now in progress.

Even if ^{Ctm}PrP is not involved in disorders other than prion diseases, the BSE epidemic highlights the importance of understanding its physiological regulation and pathophysiological cascade. Is ^{Ctm}PrP a signal transducer? If so, what is its ligand and its second messenger? Unpublished work suggests that the neurodegeneration observed in sick Tg mice overexpressing ^{Ctm}PrP is via an apoptotic mechanism (Hieronymus et al., unpublished). The precise pathway and mode of regulation of this apoptotic program needs to be determined.

Regardless of whether ^{Ctm}PrP is a specific manifestation of a novel general mechanism of biological regulation in higher eukaryotes or only of a sepcialized one used by a small set of unusual proteins, it seems fair to expect that its further exploration will reveal additional surprises.

References

Blobel G, Dobberstein B (1975a) Transfer of proteins across membranes. I. Presence of proteolytically processed and unprocessed nascent immunoglobulin light chains on membrane-bound ribosomes of murine myeloma. J Cell Biol 1975 67: 835–851.

Blobel G, Dobberstein B (1975b) Transfer of proteins across membranes. II. Reconstitution of functional rough microsomes from heterologous components. J Cell Biol, 67: 852–862

Bonifacino JS, McCarthy SA, Maguire JE, Nakayama T, Singer DS, Klausner RD, Singer A (1990) Novel post-translational regulation of TCR expression in CD4+CD8+thymocytes influenced by CD4. Nature 344: 247–251

Caras IW, Weddell GN (1989) Signal peptide for protein secretion directing glycophospholipid membrane anchor attachment. Science 243: 1196–8

DeArmond SJ, Prusiner SB (1995) Prion protein transgenes and the neuropathology in prion diseases. Brain Pathol 5: 77–89

Ellgaard L, Molinari M, Helenius A (1999) Setting the standards: quality control in the secretory pathway. Science 286: 1882–1888

Gorlich D, Rapoport TA (1993) Protein translocation into proteoliposomes reconstituted from purified components of the endoplasmic reticulum membrane. Cell 75: 615–630

Hay B, Barry RA, Lieberburg I, Prusiner SB, Lingappa VR (1987) Biogenesis and transmembrane orientation of the scrapie prion protein. Mol Cell Biol 7: 914–919

Hegde RS, Lingappa VR (1996) Sequence-specific alteration of the ribosomembrane junction exposes nascent secretory proteins to the cytosol. Cell 85: 217–228

Hegde RS, Mastrianni JA, Scott MR, DeFea KA, Tremblay P, Torchia M, DeArmond SJ, Prusiner SB, Lingappa VR (1998) A transmembrane form of the prion protein in neurodegenerative disease. Science 279: 827–34

Hegde RS, Voigt S, Lingappa VR (1998) Regulation of protein topology by transacting factors at the endoplasmic reticulum. Mol Cell 2: 85–91

Hegde RS, Tremblay P, Groth D, DeArmond SJ, Prusiner SB, Lingappa VR (1999) Transmissible and genetic prion diseases share a common pathway of neurodegeneration involving transmembrane prion protein Nature 402: 822–826

Katz FN, Lodish HF (1979). Transmembrane biogenesis of the vesicular stomatitis virus glycoprotein. J Cell Biol 80: 416–26

Katz FN, Rothman JE, Lingappa VR, Blobel G, Lodish HF (1977). Membrane assembly in vitro: synthesis, glycosylation and asymmetric insertion of a transmembrane protein. Proc Natl Acad Sci USA 74: 3278–3282

Klappa P, Stromer T, Zimmermann R, Ruddock LW, Freedman RB (1998(A pancreas-specific glycosylated protein disulphide-isomerase binds to misfolded proteins and peptides with an interaction inhibited by oestrogens. Eur J Biochem 254: 63–69

Lingappa JR, Hill RL, Wong ML, Hegde RS (1997). A multi-step, ATP-dependent pathway for assembly of human immunodeficiency virus capsids in a cell-free system. J Cell Biol 136: 567–581

Nicchitta CV, Blobel G (1993). Lumenal proteins of the mammalian endoplasmic reticulum are required to complete protein translocation. Cell 73: 989–98

Palade G (1975) Intracellular aspects of the process of protein synthesis. Science 189: 347–358

Rothman RE, Andrews DW, Calayag MC, Lingappa VR (1989). Construction of defined polytopic integral transmembrane proteins: the role of signal and stop transfer sequence permutations. J Biol Chem 263: 10470–10480

Schmidt-Rose T, Jentsch TJ (1997). Transmembrane topology of a CLC chloride channel. Proc Natl Acad Sci U.S.A. 94: 7633–7638

Sidrauski C, Chapman R, Walter P (1998) The unfolded protein response: an intracellular signalling pathway with many surprising features. Trends Cell Biol 8: 245–249

Simon SM, Blobel G (1991). A protein-conducting channel in the endoplasmic reticulum. Cell 65: 371–380

Simon K, Perara E, Lingappa VR (1987). Translocation of globin fusion proteins across the endoplasmic reticulum membrane in *Xenopus laevis* oocytes. J Cell Biol 104: 1165–1172

Skach WR, Calayag MC, Lingappa VR (1993) Evidence for an alternate model of human P-glycoprotein structure and biogenesis. J Bio Chem 268, 6903–6908

Stahl N, Borchelt DR, Hsiao K, Prusiner SB (1987) Scrapie prion protein contains a phosphatidylinositol glycolipid. Cell 51: 229–240

Walter P, Johnson AE (1994) Signal sequence recognition and protein targeting to the endoplasmic reticulum membrane. An Rev Cell Biol 10: 87–119

Ward CL, Kopito RR (1994) Intracellular turnover of cystioc fibrosis transmembrane conductance regulator. Inefficient processing and rapid degradation of wild-type and mutant proteins. J Biol Chem 269: 25710–25718

Wiertz EJ, Tortorella D, Bogyo M, Yu J, Mothes W, Jones TR, Rapoport TA, Ploegh HL (1996) Sec61-mediated transfer of a membrane protein from the endoplasmic reticulum to the proteasome for destruction. Nature 384: 432–438

The Value of Transgenic Models for the Study of Neurodegenerative Diseases

D. L. Price, P. C. Wong, A. L. Markowska, G. Thinakaran, M. K. Lee, L. J. Martin, J. Rothstein, S. S. Sisodia, and D. R. Borchelt

Introduction

The neurodegenerative diseases, including Alzheimer's disease (AD) and amyotrophic lateral sclerosis (ALS), are a heterogeneous group of age-associated, chronic illnesses that are among the most challenging and devastating diseases in medicine (Becker et al. 1998; Burright et al. 1995; Cleveland 1999; Davies et al. 1997; Goedert et al. 1998; Hardy and Gwinn-Hardy 1998; Lin et al. 1999; Mangiarini et al. 1996; Price et al. 1998b; Robitaille et al. 1999; Schilling et al. 1999; Selkoe 1999; Wong et al. 1998; Zoghbi and Orr 1999). Many of these disorders are characterized by onset in adult life, chronic progressive course, readily identifiable clinical syndromes, dysfunction/death of specific populations of neurons, and distinct cellular abnormalities often associated with the presence of intra- or extracellular malfolded proteins that, in many cases, tend to aggregate (Goedert et al. 1998; Lansbury 1999). For the most part, we do not fully understand many of the factors that influence the age of onset/course of disease, the vulnerabilities of subsets of neurons, or the mechanisms leading to neuronal degeneration in these different disorders. At present, treatments are symptomatic, with the vast majority of patients eventually becoming severely disabled and dying of intercurrent illnesses.

Although AD and ALS usually occur as sporadic illnesses, in some instances these disorders are familial and caused by mutations of specific genes. Thus, major clues as to pathogenesis of these illnesses have come from genetic studies. For example, familial AD (FAD) and familial ALS (FALS) show autosomal dominant inheritance: in these pedigrees, mutant genes encoding either the amyloid precursor protein (APP) or presenilins (PS1 or PS2) cause FAD (Price et al. 1998b; St. George-Hyslop 1999); and some cases of FALS are linked to mutations in the superoxide dismutase 1 (SOD1) gene (Andersen et al. 2000; Wong et al. 1998). The identification of these mutations has allowed investigators to use transfection and transgenic methods to introduce mutant genes into cells in both in vitro and in vivo systems. These systems have been used to examine some of the molecular and biochemical pathways leading to cellular abnormalities associated with the presence of mutant gene products. In some of the autosomal dominant genetic neurodegenerative disorders, the mutant proteins do not exhibit reductions in their normal functions, but instead acquire toxic properties themselves or participate in the formation of toxic products (Cleveland 1999; Goedert

Research and Perspectives in Alzheimer's Diseases
Beyreuther/Christen/Masters (Eds.)
Neurodegenerative Disorders
© Springer-Verlag Berlin Heidelberg 2001

et al. 1998; Hardy and Gwinn-Hardy 1998; Price et al. 1998a, b; Schilling et al. 1999; Selkoe 1999; Wong et al. 1998; Zoghbi and Orr 1999).

Although many of the biochemical steps in the pathways whereby mutant gene products, directly or indirectly, damage neural cells are not fully delineated, strategies are now available to begin to define some of the participants in these pathological events in these in vitro and in vivo model systems. Some of these pathways are being clarified by introducing or removing genes/gene products from cells in model systems. In this review, we illustrate, with selected examples drawn predominantly from our laboratory, the ways in which some of these issues are being approached using transgenic (Tg) and gene-targeted models and by mating strategies that introduce or remove specific genes.

Alzheimer's Disease

AD, the most common cause of dementia in adult life, is associated with the selective damage of brain regions and neural circuits critical for memory and cognition (Albert 1996; Mesulam 1999; Price 1999), including neurons in the neocortex, hippocampus, amygdala, basal forebrain cholinergic system, and brainstem monoaminergic nuclei (Morrison and Hof 1997; Price and Sisodia 1998; Sze et al. 1997). Dysfunction and death of neurons in these neural circuits lead to reduced numbers of generic synaptic markers, as well as transmitter-specific markers, in target fields (Francis et al. 1994; Masliah 1998; Price and Sisodia 1998; Sze et al. 1997; Whitehouse et al. 1982). Affected neurons accumulate tau and ubiquitin immunoreactivites within neurofibrillary tangles (NFTs), in cell bodies and dendrites, and in dystrophic neurons (Delacourte et al. 1998; Price and Sisodia 1998; Price and Morris 1999). The NFTs, intracellular aggregations of poorly soluble filaments, are composed principally of hyperphosphorylated isoforms of tau, a low molecular weight microtubule-binding protein (Delacourte et al. 1998; Goedert et al. 1998). In cortex, hippocampus, and amygdala of cases of AD, there are elevated levels of a $\sim$4kD Aβ peptide derived from the β-amyloid precursor protein (APP), which makes up extracellular fibrillar amyloid deposits (Morrison and Lansbury 1999; Price 1999; Price and Sisodia 1998; Selkoe 1999) within senile plaques, which are composed of amyloid surrounded by dystrophic neurites and glial cells (astrocytes and microglia; McGeer and McGeer 1998; Selkoe 1999). Eventually, neuronal loss occurs in many affected regions (Hof 1997).

Mutant Genes in FAD: APP and Presenilins

In some individuals, particularly those with early onset AD, the illness may be inherited in an autosomal dominant manner. This phenotype has been shown to be caused by mutations in three different genes: APP and Presenilin 1 and 2 (PS1 and PS2) (Hardy and Gwinn-Hardy 1998; Price et al. 1998b; Selkoe 1999).

APP. Localized to chromosome 21, this gene is alternatively spliced to mRNA encoding several APP species, which are typical type-I integral membrane glycoproteins, with: an N-terminal signal peptide; a large ectodomain with sites for N-glycosylation; an alternatively spliced Kunitz-type serine protease inhibitor (KPI) domain; an Aβ region (comprised of 28 amino acids of the ectodomain and 11–14 amino acids of the adjacent transmembrane domain); a single membrane-spanning helix; and a short cytoplasmic domain (Price et al. 1998b; Selkoe 1999). Some APP molecules are cleaved endoproteolytically within the Aβ sequence by APP "α-secretase", an activity with unusual properties (Sisodia 1992) to release the ectodomain of APP (APP^{sα}; Esch et al. 1990; Sisodia et al. 1990), including residues 1–16 of Aβ, into the culture medium or into the CSF (Selkoe 1999; Sisodia et al. 1990; Weidemann et al. 1989). The cleavage of APP within the Aβ domain precludes the formation of Aβ. In contrast, Aβ is generated by pathways involving the endoproteolytic cleavage of APP by an enzyme identified as β-secretase (BASE; Sinha et al. 1999; Vassar et al. 1999; Yan et al. 1999) and by an activity termed "γ-secretase", which in concert generate the N- and C-termini of the Aβ peptide, respectively (Hardy and Gwinn-Hardy 1998; Hardy and Israël 1999; Price et al. 1998b; Selkoe 1999). BASE has recently been cloned and characterized (Sinha et al. 1999; Vassar et al. 1999; Yan et al. 1999) and shown to be a transmembrane aspartyl protease that cleaves APP between residues 671 and 672 of APP (Sinha et al. 1999; Vassar et al. 1999; Yan et al. 1999) BASE represents a new therapeutic target. γ-secretase, which cleaves at the N-terminal portion of Aβ (Selkoe 1999), has not yet been cloned. Some investigators have suggested that PS1 (see below) is γ-secretase (i.e., PS1 is an autoactivated intramembranous aspartyl protease), whereas others believe that PS1 is a co-factor essential for secretase activity, or that it plays a key role in trafficking APP to the proper compartment for processing (Hardy and Israël 1999; Naruse et al. 1998; Wolfe et al. 1999). Although Aβ may appear in the endoplasmic reticulum or Golgi, most Aβ appears to be produced in endosomes. Thus, APP processing leads to a major secreted pool containing Aβ40 and Aβ42; a minor non-secreted intracellular pool may also exist (Wilson et al. 1999) and could act as a internal neurotoxin (Rosenblum 1999). Secreted Aβ42, when released into the neuropil, appears to damage processes of neurons (particularly nerve terminals); eventually synapses and finally neurons die (Lansbury 1999). As yet to be determined factors in the aged primate brain may contribute to the degeneration of nerve cells (Geula et al. 1998).

APP isoforms are present in many types of cells, including neurons and, at lower levels, astrocytes. In neurons, APP is transported within axons by the fast anterograde system (Koo et al. 1990; Sisodia et al. 1993); for example, in peripheral sensory neurons of rodents, APP-695 is the predominant isoform. Full-length APP-695 and, to a lesser extent, APP-751/770 are rapidly transported anterogradely in axons (Koo et al. 1990; Sisodia et al. 1993). Radiolabelling studies in the entorhinal cortex have shown that newly synthesized APP, principally APP-695, is transported via axons of the perforant pathway to accumulate at presynaptic terminals in the hippocampal formation (Buxbaum et al. 1998). In the terminal fields of the perforant pathway of rats, soluble COOH-terminally truncated

APP and amyloidogenic C-terminal fragments have been identified (Buxbaum et al. 1998), an observation consistent with the idea that neurons and their pro- cesse (i.e., axon terminals) are one source of the APP that gives rise to Aβ species. Consistent with this idea are: studies in a line of mutant APP Tg mice (Calhoun et al. 1999) showing Aβ deposits in proximity to swollen neurites, which were prelabeled with a marker injected in the entorhinal cortex and anterogradely transported to the hippocampus (Phinney et al. 1999), and investigations of mutant APP:APP −/− mice, whose APP was derived from a transgene expressed only in neurons, which showed Aβ deposits in brain and around cerebral vessels (Calhoun et al. 1999). These findings are also in accord with studies of APP Tg mice, aged monkeys, and humans with AD, showing that Aβ deposits are often located in proximity to neurites (Arnold et al. 1991; Borchelt et al. 1997; Martin et al. 1991; Walker and Cork 1999). For example, in cases of AD, Aβ deposits/neu- ritic plaques may align in the outer portion of the molecular layer of the dentate gyrus (i.e., the terminal fields of the perforant pathway whose neuronal cell bod- ies, often showing NFT, reside in the entorhinal cortex; Arnold et al. 1991).

A variety of APP mutations, including APPswe (a double mutation at the N- terminal of Aβ) and APP-717 (near the C-terminus of Aβ), have been reported in cases of FAD (Goate et al. 1991; Mullan et al. 1992). Cells that express mutant APP show aberrant APP processing: with the APPswe mutation, levels of Aβ are elevated; with the APP (717) mutation, there is a higher secreted fraction of lon- ger Aβ peptides (APP-717) relative to cells that express wild-type APP (Citron et al. 1992; Price et al. 1998b; Selkoe 1999; Suzuki et al. 1994).

PS1 and PS2. These genes, localized to chromosomes 14 (PS1) and 1 (PS2), respectively, encode highly homologous 43- to 50-kD polytopic proteins (Doan et al. 1996; Sherrington et al. 1995) that contain multiple transmembrane (TM) domains as well as a hydrophilic acidic "loop" region. The N-terminus, loop, and C-terminus of PS1 are oriented towards the cytoplasm (Doan et al. 1996). Although PS1 is synthesized as an ∼42- to 43-kD polypeptide, the preponderant PS1-related species that accumulate in vitro and in vivo are ∼27- to 28-kD N- terminal and ∼16- to 17-kD C-terminal derivatives (Lee et al. 1996; Podlisny et al. 1997; Thinakaran et al. 1996). These fragments accumulate in a 1:1 stochio- metry, are tightly regulated and saturable, and are stably associated. PS genes are widely expressed at low abundance in the central nervous system. As described above and below, PS1 influences APP processing: PS1 mutants increase levels of Aβ, and PS1 knockouts reduce levels of Aβ. It is not known whether the influence of PS is a direct effect of PS1 acting as an aspartyl protease, by its activity as a co- factor critical for the activity of γ-secretase, or by impacting on trafficking of APP (De Strooper et al. 1998; Naruse et al. 1998; Wolfe et al. 1999).

The PS1 gene has been reported to harbor >50 different FAD mutations in >80 families whereas only a small number of mutations have been found in PS2 (Cruts et al. 1998; Hardy 1997). The vast majority of the genetic abnormalities in PS genes are missense mutations that result in single amino acid substitutions; however, a mutation that deletes exon 9 from PS1 has been identified in several different FAD families.

Transgenic Models of Aβ Amyloidogenesis

To attempt to generate animal models of amyloidogenesis and Aβ-associated abnormalities, many groups have created Tg mice that express wild-type APP, APP fragments, Aβ, or FAD-linked mutant APP or PS1 transgenes. Although early efforts were disappointing because Tg mice did not exhibit any of the cellular abnormalities characteristic of AD, more recent work has shown that multiple lines of Tg mice now exist that show Aβ deposits and neuritic plaques (Borchelt et al. 1997; Calhoun et al. 1998; Holcomb et al. 1998; Hsiao et al. 1996; Irizarry et al. 1997; Masliah et al. 1996; Price et al. 1998b; Sturchler-Pierrat et al. 1997; Calhoun et al. 1999; Phinney et al. 1999).

Although a variety of groups have had success with this approach, we illustrate the strategy with studies from our laboratory. Using a recently engineered PrP vector (Borchelt et al. 1996). Dr. Borchelt and colleagues have produced two lines of mice that express the Mo/HuAPPswe transgene product at levels that are 2- to 3-fold higher than the level of endogenous Mo-APP. Levels of Aβ40 and Aβ42 in brain are increased and, by 18–20 months of age, these Mo/HuAPPswe Tg mice develop diffuse and compact Aβ immunoreactive deposits in brain (Borchelt et al. 1997). Many of the amyloid deposits are surrounded by enlarged dystrophic neurites and by glial cells. However, like most other mice overexpressing mutant APP, these mice do not show striking evidence of tau-related pathology. Preliminary cross-sectional behavioral studies on one line of mice made congenic by backcrossing to C57BL/6J mice suggest that these mice with APPswe mutations appear to develop memory deficits.

Gene Targeted Mice

Gene targeting of APP and PS1 allows examination of the consequences of ablation of specific genes. App$^{-/-}$ mice show only a subtle phenotype (Zheng et al. 1995). However, PS1$^{-/-}$ mice fail to survive beyond the early postnatal period and exhibit severe perturbations in the development of the axial skeleton and ribs (Shen et al. 1997; Wong et al. 1997); these abnormal developmental patterns are highly reminiscent of somite segmentation defects described in mice with functionally inactivated Notch1 and Delta Dll1 (a Notch ligand; Conlon et al. 1995; Hrabe de Angelis et al. 1997). Both wild-type and mutant PS1 rescue the developmental phenotype of PS–/– mice (Davis et al. 1998; Qian et al. 1998). Significantly, PS1 deficiency appears to significantly reduce the levels of Aβ in brains of the PS1$^{-/-}$ embryos (De Strooper et al. 1998; Naruse et al. 1998). Whether this outcome occurs because loss of PS1 interferes with membrane trafficking of APP and other proteins in neurons, because it is a critical cofactor in the activity of γ-secretase, or because PS1 is the secretase itself is not yet clearly established (Hardy and Israël 1999; Naruse et al. 1998; Wolfe et al. 1999). Whatever the case, PS1 influences APP processing directly or indirectly, with ablation of PS1 leading to decrements in levels of Aβ and mutant PS1 elevating levels of Aβ.

Influences of Introduction or Ablation of Specific Genes and Aβ Vaccinations

Although they do not model the full phenotype of AD, these lines of mutant APP mice generated in several laboratories represent excellent of Aβ amyloidogenesis and are highly suitable for analyses of pathogenic pathways, identification of targets, and experimental therapies, either by introduction/subtraction of specific genes or by administration of pharmacological agents (secretase inhibitors) or Aβ vaccines (Hardy and Israël 1999; Schenk et al. 1999; Sinha et al. 1999; Vassar et al. 1999; Yan et al. 1999).

To illustrate the principle that introduction of genes can influence Aβ amyloidogenesis, we briefly describe our studies of crosses between APPswe and PS1 mutant Tg mice. For example, Tg mice that coexpress mutant A246E HuPS1 and Mo/Hu-APPswe show an acceleration of the appearance of neuritic amyloid plaques in hippocampus and cortex as compared to APPswe Tg mice (Borchelt et al. 1997). Moreover, when APPswe Tg mice are mated with PS1 △E9 Tg mice, double Tg progeny show an even greater acceleration of Aβ deposition in cortex and hippocampus (Drs. Lee, Borchelt, Sisodia, Price, personal observations). Other investigators studying APPV717F: APOE −/− mice have shown that these animals exhibit reductions of levels of Aβ as compared to Tg mice carrying the APP mutant or an ApoE wild-type background (Bales et al. 1997).

With regard to Aβ immunization, this strategy reduces Aβ amyloidogenesis in APP717 mice in both prevention and treatment trials (Schenk et al. 1999; St. George-Hyslop and Westaway 1999). In the future, Tg models will continue to be of great value in a variety of testing anti-amyloidogenic strategies, including the effects of inhibition of β- and/or γ-secretase involved in production of Aβ, etc.

Amyotrophic Lateral Sclerosis

ALS, the most common adult onset motor neuron disease, manifests as weakness/muscle atrophy and, in many cases, spastic paralysis/extension plantar responses, reflecting the selective involvement of lower and upper motor neurons (Rowland 1994; Wong et al. 1998). The neuropathological features of lower motor neurons include: hyperaccumulation of phosphorylated neurofilaments in cell bodies; cytoplasmic inclusions showing ubiquitin immunoreactivity (and, in mutant SOD1 FALS cases, SOD1 immunoreactivity); cytoplasmic inclusions resembling Lewy bodies; abnormally fragmented Golgi; attenuated dendrites; neurofilamentous swellings in proximal axonal segments; and Wallerian degeneration (Carpenter 1968; Chou 1992; Hirano 1996; Ince 2000; Martin 2000; Wong et al. 1998). Recent evidence suggests that degenerating motor neurons pass through a series of stages: chromatolysis, perikaryl, dendritic atrophy, and apoptosis (Martin 2000). In the late stages, the following abnormalities relevant to apoptotic processes have been noted: some neurons show TUNEL labeling; samples of ALS spinal exhibit internucleosomal DNA fragments(laddering); the presence of DFF 40 and evidence of caspase 3 activation; and aberrant shifts in the distribution of Bax and BcL_2 in the soluble and membranous (mitochondrial) compartments (Martin 2000).

Mutant Genes in FALS: SOD1

Approximately 10 % of cases of ALS are familial, and, in almost all cases, inheritance exhibits an autosomal dominant pattern (Andersen et al. 2000; Brown 1997). Approximately 15–20 % of patients with autosomal dominant familial ALS (FALS) have missense mutations in the gene that encodes cytosolic Cu/Zn superoxide dismutase 1 (SOD1; Deng et al. 1993; Rosen et al. 1993); the enzyme that catalyzes the conversion of $\cdot O_2$ to O_2 and H_2O_2 (Fridovich 1986; Stadtman 1992). To date, >50 different missense mutations and more than one frame shift mutation have been identified in the SOD1 gene (Andersen et al. 2000; Wong and Borchelt 1995); the phenotypes associated with different mutations may show some differences, but all result in motor neuron disease (Cudkowicz et al. 1998).

In view of the extensive studies on free radical toxicity and the role of SOD1 in scavenging superoxide, investigators initially proposed that the lesions seen in SOD1-linked FALS could result from diminished free radical scavenging activity (Deng et al. 1993). However, multiple lines of evidence fail to support this hypothesis. Instead, the overwhelming evidence is entirely consistent with the concept that FALS-linked mutations cause SOD1 to acquire toxic properties. The following observations are in accord with this idea: 1) SOD1-linked FALS is an autosomal dominant disease and no null mutations have been identified; 2) assays of mutant SOD1 activity in transfected cells indicate that, although some FALS mutations reduce the activity of SOD1, others retain near-normal levels of enzyme activity/stability (Borchelt et al. 1994; Bowling et al. 1993; Yim et al. 1996); 3) mutant SOD1 subunits do not appear to alter the metabolism/ activities of wild-type SOD1 in a dominant negative fashion (Borchelt et al. 1995); 4) most FALS mutant SOD1 rescue the growth defects of SOD1-deficient yeast (Rabizaden et al. 1995), which are extremely sensitive to superoxide toxicity and fail to grow in normal atmosphere if lysine or methionine is removed from culture media; 5) the mutants show chaperone-facilitated copper binding (Corson et al. 1998; Yim et al. 1996); 6) mutant enzymes appear to exhibit enhanced ability to generate free radicals (Yim et al. 1996); 7) althoug SOD1 null mice show increased vulnerability of motor neurons to axotomy, they do not develop a FALS-like syndrome (Reaume et al. 1996); and finally, 8) multiple lines of Tg mice expressing several different FALS-linked mutant SOD1 develop many, if not all, of the clinical and pathological hallmarks of motor neuron disease (Borchelt et al. 1998; Bruijn et al. 1997b; Cleveland 1999; Dal Canto and Gurney 1994, 1997; Gurney et al. 1994; Ripps et al. 1995; Wong et al. 1995).

The molecular mechanisms by which mutant SOD1 causes degeneration of motor neurons are not fully understood. Initially, two not mutually exclusive hypotheses were proposed to explain the toxic properties acquired by the mutant

SOD1: 1) the improperly folded mutant SOD1 catalyzes, via –OONO and nitronium intermediates, the nitration of tyrosines on proteins critical for the proper functions and viability of motor neurons (Beckman et al. 1993); and 2) the mutant SOD1 possesses enhanced peroxidase activities that generate H_2O_2 and elevate levels of hydroxyl radicals (Wiedau-Pazos et al. 1996) that oxidize targets important for the normal biological functions of motor nerve cells. These hypotheses have generated many experiments and there is evidence of oxidative damage in human cases and model systems (Beal et al. 1997; Bogdanov et al. 1998; Ferrante et al. 1997; Lin et al. 1998; Pedersen et al. 1998). Nevertheless, the pathogenic mechanisms and molecular targets of damage are still uncertain. However, an emerging view, consistent with experimental results (Bruijn et al. 1997a; Cleveland 1999; Hottinger et al. 1997; Wiedau-Pazos et al. 1996; Wong et al. 1995, 2000), is that the SOD1 mutations induce conformational changes SOD1 and allow the participation of bound copper in deleterious chemistries that generate reactive molecules that can damage, perhaps via oxidation or nitration, a variety of cell constituents required for the maintenance and survival of motor neurons. Consistent with this idea is the finding that, in cases of sporadic and familial ALS, elevations occur in free nitrotyrosine and in several markers of oxidative damage (Beal et al. 1997; Ferrante et al. 1997; Pedersen et al. 1998). It should be emphasized that copper can be very toxic (Waggoner et al. 1999) and free copper is virtually absent in cytoplasm (Lippard 1999; Rae et al. 1999).

Although there is, as of yet, little experimental proof that copper bound to mutant SOD1 plays a key role in the generation of toxicity, recent work in yeast has provided information critical for designing experiments to test this hypothesis. Studies in yeast and more recently mammalian cells indicate that delivery of copper to specific proteins is mediated through distinct intracellular pathways of copper trafficking (Culotta et al. 1997; Pufahl et al. 1997; Rae et al. 1999; Valentine and Gralla 1997; Wong et al. 2000), which involve copper carriers (or metal ion chaperones). These soluble proteins, which deliver copper ions to specific intracellular proteins (Lippard 1999; Rae et al. 1999), ensure the proper intracellular transport and compartmentalization of this transition metal and help to keep the levels of free copper extraordinarily low. In this way, the potential toxicity of free ionic copper is minimized. Of particular interest is the recent work on the copper chaperone superoxide dismutase (CCS), which delivers copper to SOD1 (see above and below). It should be emphasized that the copper toxicity hypothesis in SOD1-linked FALS involves copper bound to the mutant enzyme, and toxicity is not believed to be related to excess or reduced levels of intracellular copper as occurs in Wilson's disease or Menke's disease, respectively (Harris and Gitlin 1996; Waggoner et al. 1999). Experiments that explore aspects of this hypothesis are described below.

Mutant SOD1 Tg Mice

A variety of experimental models of motor neuron disease have been described (Elliott 1999; Price et al. 1994). With regard to models, Tg mice expressing a variety of mutant SOD1 develop progressive weakness and muscle atrophy (Borchelt et al. 1998; Bruijn et al. 1997b; Chin et al. 1995; Dal Canto and Gurney 1994, 1997; Gurney 2000; Gurney et al. 1994; Kong and Xu 1997; Morrison et al. 1996, 1998; Ripps et al. 1995; Tu et al. 1996; Williamson et al. 1998; Wong et al. 1995). Pathological examinations have demonstrated SOD1/ubiquitin immunoreactive cytoplasmic aggregates with motor nerve cells; irregularly enlarged dendrites and abnormalities of motor axons (both of which show swollen mitochondria); abnormal patterns in neurofilament immunoreactivity in cell bodies/axons of motor neurons, Wallerian degeneration of motor axon; and, finally, death of these neurons.

The G37R mutant SOD1 Tg mice accumulate the G37R SOD1 to 3–12× levels of endogenous SOD1 in the spinal cord (Wong et al. 1995); the mutant SOD1 retains full specific activity. Levels of the mutant transgene product determine the age of onset. At 5–7 weeks of age (~2–3 months before the appearance of clinical signs), SOD1 accumulates in irregular swollen intraparenchymal portions of motor axons, the axonal cytoskeleton is abnormal, and vacuoles are present in these axons. Radiolabeling studies have demonstrated that both endogenous SOD1 and G37R SOD1 are transported anterogradely in axons (Borchelt et al. 1998), including degeneration of mitochondria (Kong and Xu 1997). Thus, toxic SOD1 is transported anterograde accumulates early in the disease, and is associated with early structural pathology in axons (Borchelt et al. 1998). Axonal transport is abnormal in these mutant SOD1 mice (Williamson and Cleveland 1999; Zhang et al. 1997). Dendrites are also abnormal, showing small vacuoles (Wong et al. 1995), some of which appear to originate in the space between the outer and inner mitochondrial membranes; eventually the mitochondrial membrane causes damage and there is disruption of the cristae. The dendritic abnormalities are reminiscent of changes seen in excitotoxicity, which has been suggested to play a role in ALS (Jackson and Rothstein 2000; Leigh and Meldrum 1996; Rothstein 1996; Shaw 2000). The cell bodies of some neurons showed SOD1 and ubiquitin-immunoreactive inclusions and phosphorylated NF-H immunoreactivities. Axonal and dendritic abnormalities, as well as intracellular aggregates, occur prior to the onset of clinical signs. However, once Wallerian degeneration is obvious, the mice usually have clinical signs. Eventually, the number of motor neurons is reduced, and astrocytes are increased in ventral horns.

Lines of mutant SOD1 mice have been used for pharmacological and "genetic" therapeutic trials (Cleveland 1999; Couillard-Després et al. 1998; Friedlander et al. 1997; Julien 1999; Klivenyi et al. 1999; Pasinelli et al. 1998).

The G93A SOD1 Tg mice (Gurney 2000) have been widely used to test a variety of therapeutic approaches. Vitamin E and selenium modestly delay both the onset and progression of disease without affecting survival. In contrast, riluzole and gabapentin do not influence the onset/progression but do increase survival slightly (Gurney et al. 1996). Oral administration of d-penicillamine, a copper chelator, delays the onset of disease (Hottinger et al. 1997). At present, treatment with creatine appears to have the most robust pharmacological impact on disease (Klivenyi et al. 1999). Overexpression of Bcl-2 in these Tg mice extends their survival, but the presence of the gene does not change the progression of the disease (Kostic et al. 1997). In G93A SOD1 mice overexpressing a dominant negative inhibitor of ICE (interleukin-1B converting enzyme), a cell death gene, there is a modest slowing of progression of disease (Friedlander et al. 1997).

Lines of mice expressing mutant G37R or G85R SOD1 have been used to examine the influence of other genes. For example, to test the role of neurofilaments in motor neuron disease caused by SOD1 mutations, G37R SOD1 Tg mice were crossbred to: 1) Tg mice that accumulate NF-H-β-galactosidase fusion protein (NF-H-lacZ), a multivalent protein that crosslinks neurofilaments in neuronal perikarya and limits their export to axons (Eyer and Peterson 1994); and 2) Tg mice expressing human NF-H subunits (Côté et al. 1993). In G37R SOD1 mice expressing NF-H-lacZ, NF are withheld from the axonal compartment, but there is no influence on disease (Eyer et al. 1998), implying that neither initiation nor progression of pathology requires an axonal NF cytoskeleton and that alterations in NF biology observed in some forms of motor neuron disease may be secondary responses (Eyer et al. 1998). By contrast, the expression of wt type human NF-H transgenes in the SOD1 mutant mice increased the mean life span of the G37R SOD1 mice. The singly G37R and the compound G37R:NF-H Tg mice had mean life expectancies of 9.5 +/- 2.8 months and 15.8 +/- 1.5 months, respectively (Couillard-Després et al. 1998). In contrast to the striking axonal degeneration observed in one-year-old Tg mice expressing G37R SOD1, the compound G37R:NF-H Tg littermates showed sparing of motor neurons (Couillard-Després et al. 1998; Julien 1999). Similarly, G85R SOD1:NF-L –/– mice showed later onset and a slower progression (Julien 1999; Williamson et al. 1998). In NF-L –/–, levels of NF-M and NF-H are reduced in axons (but not cell bodies). The common property shared by the G37R SOD1:NF-H Tg mice and the 685R SOD1:NF-L Tg mice is the reduced content of assembled neurofilaments in the axonal compartment, an observation that lends credence to the idea that, in some way, the presence of neurofilaments in axons renders the motor neuron vulnerable to the toxic properties of mutant SOD1.

The effects of increasing or reducing the levels of SOD1 have been tested by mating mutant SOD1 mice with Tg mice overexpressing wt SOD1 or by mating them with SOD1 –/– mice. Neither of these genetic manipulations appeared to have a significant effect on the phenotype of mutant SOD1 mice in these experiments (Bruijn et al. 1998; i.e., the clinical course and pathology, including the presence of aggregates, which contain SOD1 and other unidentified components, are no different from those of mutant SOD1 Tg mice).

Gene Targeting of SOD1 and CCS

SOD1–/– mice do not develop a FALS syndrome, but motor neurons in these animals are more vulnerable to axotomy (Reaume et al. 1996). As indicated above, crossing SOD1 mutant mice with SOD1–/– mice does not influence the course of disease in the progeny (Bruijn et al. 1998).

If mutant SOD1 bound copper plays a role in toxicity, it would be of great interest to reduce the levels of copper in SOD1. Until recently, this has not been possible. However, the discovery of CCS as a copper chaperone that delivers the metal to SOD1 has allowed investigators to begin to test this hypothesis. CCS, whose protein domain/crystal structure has been determined (Lamb et al. 1999; Schmidt et al. 1999), has been shown to interact with both wild-type and FALS-linked mutant SOD1, and copper incorporation into these molecules is CCS dependent (Rae et al. 1999; Wong et al. 2000). Thus, CCS delivers copper to SOD1. Moreover, CCS is present in the nervous system, including upper and lower motor neurons (Rothstein et al. 1999). Dr. P. Wong and colleagues have gene targeted CCS, and the phenotype in the CCS null mice, not surprisingly, resembles that of the SOD1$^{-/-}$ mice (Wong et al. 2000). The mating of CCS$^{-/-}$ mice to mutant SOD1 mice will allow testing of the hypotheses that copper plays a role in the toxicity of mutant SOD1. Demonstrating that mutant SOD1 Tg mice lacking CCS show significant amelioration of disease in spite of increased levels of mutant SOD1 would strengthen the argument that aberrant copper chemistries play important roles in the pathogenesis of disease in mutant SOD1 Tg mice. This outcome would greatly encourage investigators to design approaches to intervene in the CCS-SOD1 copper trafficking pathway as therapy for individuals with SOD1-linked FALS and, potentially, for other forms of motor neuron disease.

Conclusions

Transgenic strategies have allowed investigators to produce mice that recapitulate some, if not all, of the features of APP- and PS1-linked FAD and of SOD1-linked FALS (Cleveland 1999; Price et al. 1998a, b; Wong et al. 1998), as well as aspects of other neurodegenerative diseases (Lee and Trojanowski 1999; Lin et al. 1999; Schilling et al. 1999; Tremblay et al. 2000). These models are proving to be very useful in investigations designed to: test the roles of specific genes in disease; define the nature of the cellular/biochemical/molecular alterations in neural tissues; delineate the character and evolution of a variety of pathologies, particularly those neuronal and/or glial abnormalities associated with protein folding/ aggregation/fibrillogenesis and apoptosis; clarify the mechanisms by which the mutant proteins cause damage to specific populations of neurons; and explore the biochemical pathways associated with dysfunction/death of nerve cells.

The emerging view is that each of the genetic forms of these diseases results because of the presence of the mutant proteins, in some instances improperly folded or aggregated, which directly or indirectly trigger pathogenic cascades that eventually impact the structure and function of subsets of neural cells. Ultimately, some of the affected nerve cells die, in some cases by apoptotic mechanisms. Complemented by in vitro investigations, future studies of Tg animals, particularly those animals carrying inducible genes (Gingrich and Roder 1998; Mansuy et al. 1998; Tremblay et al. 2000), will be of great value in further defining some of the in vivo events occurring in models of these illnesses. Moreover, new technologies, including laser capture microscopy, array technologies, and cell biological/structural biophysical methods, will help investigators to define the changes in levels of a variety of genes/gene products as the disease evolves, the conformational changes in mutant proteins and their impact on biological properties of these proteins, and the biochemical events that are implicated in these pathogenic cascades (Fend et al. 1999; Fink et al. 1998; Lamb et al. 1999; Lin et al. 1999; Luo et al. 1999; Simone et al. 1998). Much of this information will be very helpful in addressing similar issues via the analyses of tissues from humans with the illnesses. Inturn, the information derived from Tg mice and humans will provide insights into potential therapeutic targets that can be further evaluated by designing novel treatments for testing in Tg models. The demonstration of efficacies in these model systems should then be rapidly translated into new treatments for these devastating human neurodegenerative disorders.

Acknowledgments

The authors wish to thank our many colleagues at JHMI and other institutions for their contributions to some of the original work cited in this review and for helpful discussions. Aspects of this work were supported by grants from the U.S. Public Health Service (AG05146, AG07914, AG14248, NS07435, NS37145, NS10580) as well as the Metropolitan Life Foundation, American Health Assistance Foundation, Hereditary Disease Foundation, Adler Foundation, and Bristol-Myers Squibb.

References

Albert MS (1996) Cognitive and neurobiologic markers of early Alzheimer disease. Proc Natl Acad Sci USA 93: 13547–13551

Andersen PM, Morita M, Brown RH Jr (2000) Genetics of amyotrophic lateral sclerosis: an overview. In: Brown RH Jr, Meininger V, Swash M (eds) Amyotrophic lateral sclerosis. London, Martin Dunitz Ltd., 223–250

Arnold SE, Hyman BT, Flory J, Damasio AR, Van Hoesen GW (1991) The topographical and neuroanatomical distribution of neurofibrillary tangles and neuritic plaques in the cerebral cortex of patients with Alzheimer's disease. Cereb Cortex 1: 103–116

Bales KR, Verina T, Dodel RC, Du Y, Altstiel M, Bender M, Hyslop P, Johnstone EM, Little SP, Cummins DJ, Piccardo P, Ghetti B, Paul SM (1997) Lack of apolipoprotein E dramatically reduces amyloid B-peptide deposition. Nature Genet 17: 263–264

Beal MF, Ferrante RJ, Browne SE, Matthews RT, Kowall NW, Brown RH Jr. (1997) Increased 3-nitrotyrosine in both sporadic and familial amyotrophic lateral sclerosis. Ann Neurol 42: 646–654

Becher MW, Kotzuk JA, Sharp AH, Davies SW, Bates GP, Price DL, Ross CA (1998) Intranuclear neuronal inclusions in Huntington's disease and dentatorubral and pallidoluysian atrophy: correlation between the density of inclusions and IT15 CAG triplet repeat length. Neurobiol Dis 4: 387–397

Beckman JS, Carson M, Smith CD, Koppenoll WH (1993) ALS, SOD and peroxynitrite Nature 364: 584

Bogdanov MB, Ramos LE, Xu Z, Beal MF (1998) Elevated "hydroxyl radical" generation in vivo in an animal model of amyotrophic lateral sclerosis. J Neurochem 71: 1321–1324

Borchelt DR, Lee MK, Slunt HH, Guarnieri M, Xu Z-S, Wong PC, Brown RH Jr, Price DL, Sisodia SS, Cleveland DW (1994) Superoxide dismutase 1 with mutations linked to familial amyotrophic lateral sclerosis possesses significant activity. Proc Ntl Acad Sci USA 91: 8292–8296

Borchelt DR, Guarnieri M, Wong PC, Lee MK, Slunt HS, Xu Z-S, Sisodia SS, Price DL, Cleveland DW (1995) Superoxide dismutase 1 subunits with mutations linked to familial amyotrophic lateral sclerosis do not affect wild-type subunit function. J Biol Chem 270: 3234–3238

Borchelt DR, Davis J, Fischer M, Lee MK, Slunt HH, Ratovitsky T, Regard J, Copeland NG, Jenkins NA, Sisodia SS, Price DL (1996) A vector for expressing foreign genes in the brains and hearts of transgenic mice. Genet Anal (Biomed Eng) 13: 159–163

Borchelt DR, Ratovitski T, Van Lare J, Lee MK, Gonzales VB, Jenkins NA, Copeland NG, Price DL, Sisodia SS (1997) Accelerated amyloid deposition in the brains of transgenic mice co-expressing mutant presenilin 1 and amyloid precursor proteins. Neuron 19: 939–945

Borchelt DR, Wong PC, Becher MW, Pardo CA, Lee MK, Xu Z-S, Thinakaran G, Jenkins NA, Copeland NG, Sisodia SS, Cleveland DW, Price DL, Hoffman PN (1998) Axonal transport of mutant superoxide dismutase 1 and focal axonal abnormalities in the proximal axons of transgenic mice. Neurobiol Dis 5: 27–35

Bowling AC, Schulz JB, Brown RH Jr, Beal MF (1993) Superoxide dismutase activity, oxidative damage, and mitochondrial energy metabolism in familial and sporadic amyotrophic lateral sclerosis. J Neurochem 61: 2322–2325

Brown RH Jr (1997) Amyotrophic lateral sclerosis. Insights from genetics. Arch Neurol 54: 1246–1250

Bruijn LI, Beal MF, Becher MW, Schulz JB, Wong PC, Price DL, Cleveland DW (1997a) Elevated free nitrotyrosine levels, but not protein-bound nitrotyrosine or hydroxyl radicals, throughout amyotrophic lateral sclerosis (ALS)-like disease implicate tyrosine nitration as an aberrant in vivo property of one familial ALS-linked superoxide dismutase 1 mutant. Proc Natl Acad Sci USA 94: 7606–7611

Bruijn LI, Becher MW, Lee MK, Anderson KL, Jenkins NA, Copeland NG, Sisodia SS, Rothstein JD, Borchelt DR, Price DL, Cleveland DW (1997b) ALS-linked SOD1 mutant G85R mediates damage to astrocytes and promotes rapidly progressive disease with SOD1-containing inclusions. Neuron 18: 327–338

Bruijn LI, Houseweart MK, Kato S, Anderson KL, Anderson SD, Ohama E, Reaume AG, Scott RW, Cleveland DW (1998) Aggregation and motor neuron toxicity of an ALS-linked SOD1 mutant independent from wild-type SOD1. Science 281: 1851–1854

Burright EN, Clark HB, Servadio A, Matilla T, Feddersen RM, Yunis WS, Duvick LA, Zoghbi HY, Orr HT (1995) SCA1 transgenic mice: a model for neurodegeneration caused by an expanded CAG trinucleotide repeat. Cell 82: 937–948

Buxbaum JD, Thinakaran G, Koliatsos V, O'Callahan J, Slunt HH, Price DL, Sisodia SS (1998) Alzheimer amyloid protein precursor in the rat hippocampus: transport and processing through the perforant path. J Neurosci 18: 9629–9637

Calhoun ME, Wiederhold K-H, Abramowski D, Phinney AL, Probst A, Stuchler-Pierrat C, Staufenbiel M, Sommer B, Jucker M (1998) Neuron loss in APP transgenic mice. Nature 395: 755–756

Calhoun ME, Burgermeister P, Phinney AL, Stalder M, Tolnay M, Wiederhold K-H, Abramowski D, Sturchler-Pierrat C, Sommer B, Staufenbiel M, Jucker M (1999) Neuronal overexpression of mutant amyloid precursor protein results in prominent deposition of cerebrovascular amyloid. Proc Natl Acad Sci USA 96: 14088–14093

Carpenter S (1968) Proximal axonal enlargement in motor neuron disease. Neurology 18: 841–851

Chiu AY, Zhai P, Dal Canto MC, Peters TM, Kwon YW, Prattis SM, Gurney ME (1995) Age-dependent penetrance of disease in a transgenic mouse model of familial amyotrophic lateral sclerosis. Mol Cell Neurosci 6: 349–362

Chou SM (1992) Pathology-light microscopy of amyotrophic lateral sclerosis. In: Smith RA (ed) Handbook of amyotrophic lateral sclerosis. New York, Marcel Dekker, pp. 133–181

Citron M, Oltersdorf T, Haass C, McConlogue L, Hung AY, Seubert P, Vigo-Pelfrey C, Lieberburg I, Selkoe DJ (1992) Mutation of the β-amyloid precursor protein in familial Alzheimer's disease increases β-protein production. Nature 360: 672–674

Cleveland DW (1999) From charcot to SOD1: mechanisms of selective motor neuron death in ALS. Neuron 24: 515–520

Conlon RA, Reaume AG, Rossant J (1995) *Notch 1* is required for the coordinate segmentation of somites. Development 121: 1533–1545

Corson LB, Culotta VC, Cleveland DW (1998) Chaperone-facilitated copper binding is a property common to several classes of familial amyotrophic lateral sclerosis-linked superoxide dismutase mutants. Proc Natl Acad Sci USA 95: 6361–6366

Couillard-Després S, Zhu Q, Wong PC, Price DL, Cleveland DW, Julien J-P (1998) Protective effect of neurofilament NF-H overexpression in motor neuron disease induced by mutant superoxide dismutase. Proc Natl Acad Sci USA 95: 9626–9630

Côté F, Collard J-F, Julien J-P (1993) Progressive neuronopathy in transgenic mice expressing the human neurofilament heavy gene: a mouse model of amyotrophic lateral sclerosis. Cell 73: 35–46

Cruts M, van Duijn CM, Backhovens H, van den Broeck M, Wehnert A, Serneels S, Sherrington R, Hutton M, Hardy J, St George-Hyslop PH, Hofman A, Van Broeckhoven C (1998) Estimation of the genetic contribution of presenilin-1 and -2 mutations in a population-based study of presenile Alzheimer disease. Human Mol Genet 71: 43–51

Cudkowicz MF, McKenna-Yasek D, Chen C, Hedley-Whyte ET, Brown RH Jr (1998) Limited corticospinal tract involvement in amyotrophic lateral sclerosis subjects with the A4V mutation in the copper/zinc superoxide dismutase gene. Ann Neurol 43: 703–710

Culotta VC, Klomp LWJ, Strain J, Casareno RLB, Krems B, Gitlin GD (1997) The copper chaperone for superoxide dismutase. J Biol Chem 272: 23469–23472

Dal Canto MC, Gurney ME (1994) Development of central nervous system pathology in a murine transgenic model of human amyotrophic lateral sclerosis. Am J Pathol 145: 1271–1280

Dal Canto MC, Gurney ME (1997) A low expressor line of transgenic mice carrying a mutant human Cu,Zn superoxide dismutase (SOD1) gene develops pathological changes that most closely resemble those in human amyotrophic lateral sclerosis. Acta Neuropathol 93: 537–550

Davies SW, Turmaine M, Cozens BA, DiFiglia M, Sharp AH, Ross CA, Scherzinger E, Wanker EE, Mangiarini L, Bates GP (1997) Formation of neuronal intranuclear inclusions underlies the neurological dysfunction in mice transgenic for the HD mutation. Cell 90: 537–548

Davis JA, Naruse S, Chen H, Eckman C, Younkin S, Price DL, Borchelt DR, Sisodia SS, Wong PC (1998) An Alzheimer's disease-linked PS1 variant rescues the developmental abnormalities of PS1-deficient embryos. Neuron 20: 603–609

Delacourte A, David JP, Sergeant N, Buee L, Wattez A, Vermersch P, Ghozali F, Fallet-Biano C, Pasquier F, Lebert F, Petit H, Di Menza C (1998) The biochemical pathway of neurofibrillary degeneration in aging and Alzheimer's disease. Neurology 52: 1158–1165

De Strooper B, Saftig P, Craessaerts K, Vanderstichele H, Guhde G, Annaert W, Von Figura K, Van Leuven F (1998) Deficiency of presenilin-1 inhibits the normal cleavage of amyloid precursor protein. Nature 391: 387–390

Deng H-X, Hentati A, Tainer JA, Iqbal Z, Cayabyab A, Hung W-Y, Getzoff ED, Hu P, Herzfeldt B, Roos RP, Warner C, Deng G, Soriano E, Smyth C, Parge HE, Ahmed A, Roses AD, Hallewell RA, Pericak-Vance MA, Siddique T (1993) Amyotrophic lateral sclerosis and structural defects in Cu,Zn superoxide dismutase. Science 261: 1047–1051

Doan A, Thinakaran G, Borchelt DR, Slunt HH, Ratovitsky T, Podlisny M, Selkoe DJ, Seeger M, Gandy SE, Price DL, Sisodia SS (1996) Protein topology of presenilin 1. Neuron 17: 1023–1030

Elliott JL (1999) Experimental models of amyotrophic lateral sclerosis. Neurobiol Dis 6: 310–320

Esch FS, Keim PS, Beattie EC, Blacher RW, Culwell AR, Oltersdorf T, McClure D, Ward PJ (1990) Cleavage of amyloid β peptide during constitutive processing of its precursor. Science 248: 1122–1124

Eyer J, Peterson A (1994) Neurofilament-deficient axons and perikaryal aggregates in viable transgenic mice expressing a neurofilament-β-galactosidase fusion protein. Neuron 12: 389–405

Eyer J, Cleveland DW, Wong PC, Peterson AC (1998) Pathogenesis of two axonopathies does not require axonal neurofilaments. Nature 391: 584–587

Fend F, Emmert-Buck MR, Chuaqui R, Cole K, Lee J, Liotta LA, Raffeld M (1999) Immuno-LCM: laser capture microdissection of immunostained frozen sections for mRNA analysis. Am J Pathol 154: 61–66

Ferrante RJ, Browne SE, Shinobu LA, Bowling AC, Baik MJ, MacGarvey U, Kowall NW, Brown RH Jr, Beal MF (1997) Evidenze of increased oxidative damage in both sporadic and familial amyotrophic lateral sclerosis. J Neurochem 69: 2064–2074

Fink L, Seeger W, Ermert L, Hänze J, Stahl U, Grimminger F, Kummer W, Bohle RM (1998) Real-time quantitative RT-PCR after laser-assisted cell picking. Nature Med 4: 1329–1333

Francis PT, Cross AJ, Bowen DM (1994) Neurotransmitters and neuropeptides. In: Terry RD, Katzman R, Bick KL (eds) Alzheimer disease. New York, Raven Press, pp. 247–261

Fridovich I (1986) Superoxide dismutases. Adv Enzymol Relat Areas Mol Biol 58: 61–97

Friedlander RM, Brown RH, Gagliardini V, Wang J, Yuan J (1997) Inhibition of ICE slows ALS in mice. Nature 388: 31–31

Geula C, Wu C-K, Saroff D, Lorenzo A, Yuan M, Yankner BA (1998) Aging renders the brain vulnerable to amyloid β-protein neurotoxicity. Nature Med 4: 827–835

Gingrich JR, Roder J (1998) Inducible gene expression in the nervous system of transgenic mice. Annu Rev Neurosci 21: 377–405

Goate A, Chartier-Harlin M-C, Mullan M, Brown J, Crawford F, Fidani L, Giuffra L, Haynes A, Irving N, James L, Mant R, Newton P, Rooke K, Roques P, Talbot C, Pericak-Vance M, Roses A, Willamson R, Rossor M, Owen M, Hardy J (1991) Segregation of a missense mutation in the amyloid precursor protein gene with familial Alzheimer's disease. Nature 349: 704–706

Goedert M, Spillantini MG, Davies SW (1998) Filamentous nerve cell inclusions in neurodegenerative diseases. Curr Opin Neurobiol 8: 619–632

Gurney ME (2000) Transgenic animal models of amyotrophic lateral sclerosis. In: Brown RH Jr, Meininger V, Swash M (eds) Amyotrophic lateral sclerosis. London, Martin Dunitz Ltd. pp. 251–262

Gurney ME, Pu H, Chiu AY, Dal Canto MC, Polchow CY, Alexander DD, Caliendo J, Hentati A, Kwon YW, Deng H-X, Chen W, Zhai P, Sufit RL, Siddique T (1994) Motor neuron degeneration in mice that express a human Cu,Zn superoxide dismutase mutation. Science 264: 1772–1775

Gurney ME, Cuttings FB, Zhai P, Doble A, Taylor CP, Andrus PK, Hall ED (1996) Benefit of vitamin E, riluzole, and gabapentin in a transgenic model of familial amyotrophic lateral sclerosis. Ann Neurol 39: 147–157

Hardy J (1997) Amyloid, the presenilins and Alzheimer's disease. Trends Neurosci 20: 154–159

Hardy J, Gwinn-Hardy K (1998) Genetic classification of primary neurodegenerative disease. Science 282: 1075–1079

Hardy J, Israël A (1999) In search of γ-secretase. Nature 398: 466–467

Harris ZI, Gitlin JD (1996) Genetic and molecular basis for copper toxicity. Am J Clin Nutr 63: 836S–841S

Hirano A (1996) Neuropathology of ALS: an overview. Neurology 47: S63–S66

Holcomb L, Gordon MN, McGowan E, Yu X, Benkovic S, Jantzen P, Wright K, Saad I, Mueller R, Morgan D, Sanders S, Zehr C, O'Campo K, Hardy J, Prada C-M, Eckman C, Younkin S, Hsiao K, Duff K (1998) Accelerated Alzheimer-type phenotype in transgenic mice carrying both mutant amyloid precursor protein and presenilin 1 transgenes. Nature Med 4: 97–100

Hottinger AF, Fine EG, Gurney ME, Zurn AD, Aebischer P (1997) The copper chelator d-penicillamine delays onset of disease and extends survival in a transgenic mouse model of familial amyotrophic lateral sclerosis. Eur J Neurosci 9: 1548–1551

Hrabe de Angelis M, McIntyre J, Gossler A (1997) Maintenance of somite borders in mice requires the *Delta* homologue *Dll1*. Nature 386: 717–721

Hsiao K, Chapman P, Nilsen S, Eckman C, Harigaya Y, Younkin S, Yang F, Cole G (1996) Correlative memory deficts, Aβ elevation and amyloid plaques in transgenic mice. Science 274: 99–102

Ince PG (2000) Neuropathology. In: Brown RH Jr, Meininger V, Swash M (eds) Amyoptrophic lateral sclerosis. London, Martin Dunitz Ltd pp. 83–112

Irizarry MC, McNamara M, Fedorchak K, Hsiao K, Hyman BT (1997) App$_{Sw}$ transgenic mice develop age-related Aβ deposits and neuropil abnormalities, but no neuronal loss in CA1. J Neuropathol Exp Neurol 56: 965–973

Jackson M, Rothstein JD (2000) Excitotoxicity in amyotrophic lateral sclerosis. In: Brown RH Jr, Meininger V, Swash M (eds) Amyoptrophic lateral sclerosis. London, Martin Dunitz Ltd pp. 263–278

Julien J-P (1999) Neurofilament functions in health and disease. Curr Opin Neurobiol 9: 554–560

Klivenyi P, Ferrante RJ, Matthews RT, Bogdanov MB, Klein AM, Andreassen OA, Mueller G, Wermer M, Kaddurah-Daouk R, Beal MF (1999) Neuroprotective effects of creatine in a transgenic animal model of amyotrophic lateral sclerosis. Nature Med 5: 347–350

Kong J, Xu Z (1997) Massive mitochondrial degeneration in motor neurons triggers the onset of amyotrophic lateral sclerosis in mice expressing a mutat SOD1. J Neurosci 18: 3241–3250

Koo EH, Sisodia SS, Archer DR, Martin LJ, Weidemann A, Beyreuther K, Fischer P, Masters CL, Price DL (1990) Precursor of amyloid protein in Alzheimer disease undergoes fast anterograde axonal transport. Proc Natl Acad Sci USA 87: 1561–1565

Kostic V, Jackson-Lewis V, de Bilbao F, Dubois-Dauphin M, Przedborski S (1997) Bcl-2: prolonging life in a transgenic mouse model of familial amyotrophic lateral sclerosis. Science 277: 559–562

Lamb AL, Wernimont AK, Pufahl RA, Culotta VC, O'Halloran TV, Rosenzweig AC (1999) Crystal structure of the copper chaperone for superoxide dismutase. Nature Struct Biol 6: 1

Lansbury PT (1999) Evolution of amyloid: what normal protein folding may tell us about fibrillogenesis and disease. Proc Natl Acad Sci USA 96: 3342–3344

Lee MK, Slunt HH, Martin LJ, Thinakaran G, Kim G, Gandy SE, Seeger M, Koo E, Price DL, Sisodia SS (1996) Expression of presenilin 1 and 2 (PS1 and PS2) in human and murine tissues. J Neurosci 16: 7513–7525

Lee VMY, Trojanowski JQ (1999) Neurodegenerative taupathies: human disease and transgenic mouse models. Neuron 24: 507–510

Leigh PN, Meldrum BS (1996) Excitotoxicity in ALS. Neurology 47: S221–S227

Lin X, Cummings CJ, Zoghbi HY (1999) Expanding our understanding of polyglutamine diseases through mouse models. Neuron 24: 499–502

Lippard SJ (1999) Free copper ions in the cell? Science 284: 748–749

Liu R, Althaus JS, Ellerbrock BR, Becker DA, Gurney ME (1998) Enhanced oxygen radical production in a transgenic mouse model of familial amyotrophic lateral sclerosis. Ann Neurol 44: 763–770

Luo L, Salunga RC, Guo H, Bittner A, Joy KC, Galindo JE, Xiao H, Rogers KE, Wan JS, Jackson MR, Erlander MG (1999) Gene expression profiles of laser-capture adjacent neuronal subtypes. Nature Med 5: 117–122

Mangiarini L, Sathasivam K, Seller M, Cozens B, Harper A, Hetherington C, Lawton M, Trottier Y, Lehrach H, Davies SW, Bates GP (1996) Exon 1 of the HD gene with an expanded CAG repeat is sufficient to cause a progressive neurological phenotype in transgenic mice. Cell 87: 493–506

Mansuy IM, Winder DG, Moallem TM, Osman M, Mayford M, Hawkins RD, Kandel ER (1998) Inducible and reversible gene expression with the rtTA system for the study of memory. Neuron 21: 257–265

Martin LJ (2000) Neuronal death in amyotrophic lateral sclerosis is apoptosis: possible contribution of a programmed cell death mechanism. J Neuropathol Exp Neurol 58: 459–471

Martin LJ, Sisodia SS, Koo EH, Cork LC, Dellovade TL, Weidemann A, Beyreuther K, Masters C, Price DL (1991) Amyloid precursor protein in aged nonhuman primates. Proc Natl Acad Sci USA 88: 1461–1465

Masliah E (1998) The role of synaptic proteins in neurodegenerative disorders. Neurosci News 1: 14–20

Masliah E, Sisk A, Mallory M, Mucke L, Schenk D, Games D (1996) Comparison of neurodegenerative pathology in transgenic mice overexpressing V717F β-amyloid precursor protein and Alzheimer's disease. J Neurosci 16: 5795–5811

McGeer EG, McGeer PL (19998) Inflammation in the brain in Alzheimer's disease: implications for therapy. Neurosci News 1: 29–35

Mesulam M-M (1999) Neuroplasticity failure in Alzheimer's disease: bridging the gap between plaques and tangles. Neuron 24: 521–529

Morrison BM, Gordon JW, Ripps ME, Morrison JH (1996) Quantitative immunocytochemical analysis of the spinal cord in G86R superoxide dismutase transgenic mice: neurochemical correlates of selective vulnerability. J Comp Neurol 373: 619–631

Morrison BM, Janssen WG, Gordon JW, Morrison JH (1998) Time course of neuropathology in the spinal cord of G86R superoxide dismutase transgenic mice. J Comp Neurol 391: 64–77

Morrison JH, Hof PR (1997) Life and death of neurons in the aging brain. Science 278: 412–419

Mullan M, Crawford F, Axelman K, Houlden H, Lillius L, Winblad B, Lannfelt L (1992) A pathogenic mutation for probable Alzheimer's disease in the APP gene at the N-terminus of β-amyloid. Nature Genet 1: 345–347

Naruse S, Thinakaran G, Luo J-J, Kusiak JW, Tomita T, Iwatsubo T, Qian X, Ginty DD, Price DL, Borchelt DR, Wong PC, Sisodia SS (1998) Effects of PS1 deficiency on membrane protein trafficking in neurons. Neuron 21: 1213–1221

Pasinelli P, Borchelt DR, Houseweart MK, Cleveland DW, Brown RH (1998) Caspase-1 is activated in neural cells and tissue with amyotrophic lateral sclerosis-associated mutations in copper-zinc superoxide dismutase. Neurobiology 95: 15763–15768

Pedersen WA, Fu W, Keller JN, Markesbery WR, Appel S, Smith RG, Kasarskis E, Mattson MP (1998) Protein modification by the lipid peroxidation product 4-hydroxynonenal in the spinal cords of amyotrophic lateral sclerosis patients. Ann Neurol 44: 819–824

Phinney AL, Deller T, Stalder M, Calhoun ME, Frotscher M, Sommer B, Staufenbiel M, Jucker M (1999) Cerebral amyloid induces aberrant axonal sprouting and ectopic terminal formation in amyloid precursor protein transgenic mice. J Neurosci 19: 8552–8559

Podlisny MB, Citron M, Amarante P, Sherrington R, Xia W, Zhang J, Diehl T, Levesque G, Fraser P, Haass C, Koo EHM, Seubert P, St. George-Hyslop P, Teplow DB, Selkoe DJ (1997) Presenilin proteins undergo heterogeneous endoproteolysis between Thr_{291} and Ala_{299} and occur as stable N- and C-terminal fragments in normal and Alzheimer brain tissue. Neurobiol Dis 3: 325–337

Price DL (1999) Aging of the brain and dementia of the alzheimer type. In: Kandel ER, Schwartz JH, Jessell TM (eds) Principles of neural science. New York, McGraw-Hill, pp. 1149–1161

Price DL, Cleveland DW, Koliatsos VE (1994) Motor neurone disease and animal models. Neurobiol Dis 1: 3–11

Price DL, Sisodia SS (1998) Mutant genes in familial Alzheimer's disease and transgenic models. Annu Rev Neurosci 21: 479–505

Price DL, Sisodia SS, Borchelt DR (1998a) Genetic neurodegenerative diseases: the human illness and transgenic models. Science 282: 1079–1083

Price DL, Tanzi RE, Borchelt DR, Sisodia SS (1998b) Alzheimer's disease: genetic studies and transgenic models. Annu Rev Genet 32: 461–493

Price JL, Morris JC (1999) Tangles and plaques in nondemented aging and "preclinical" Alzheimer's disease. Ann Neurol 45: 358–368

Pufahl RA, Singer CP, Peariso KL, Lin S-J, Schmidt PJ, Fahrni CJ, Culotta VC, Penner-Hahn JE, O'Halloran TV (1997) Metal ion chaperone function of the soluble Cu(I) receptor Atx1. Science 278: 853–856

Qian S, Jiang P, Guan X, Singh G, Trumbauer M, Yu H, Chen H, van der Ploeg L, Zheng H (1998) Mutant human presenilin 1 protects presenilin 1 null mouse against embryonic lethality and elevates Aβ1-42/43 expression. Neuron 20: 611–617

Rabizadeh S, Butler Gralla E, Borchelt DR, Gwinn R, Selverston Valentine J, Sisodia SS, Wong P, Lee M, Hahn H, Bredesen DE (1995) Mutations associated with amyotrophic lateral sclerosis convert superoxide dismutase from an antiapoptotic gene to a proapoptotic gene: studies in yeast and neural cells. Proc Natl Acad Sci USA 92: 3024–3028

Rae TD, Schmidt PJ, Pufahl RA, Culotta VC, O'Holloran TV (1999) Undetectable intracellular free copper: the requirement of a copper chaperone for superoxide dismutase. Science 284: 805–808

Reaume AG, Elliott JL, Hoffman EK, Kowall NW, Ferrante RJ, Siwek DF, Wilcox HM, Flood DG, Beal MF, Brown RH Jr, Scott RW, Snider WD (1996) Motor neurons in Cu/Zn superoxide dismutase-deficient mice develop normally but exhibit enhanced cell death after axonal injury. Nature Genet 13: 43–47

Ripps ME, Huntley GW, Hof PR, Morrison JH, Gordon JW (1995) Transgenic mice expressing an altered murine superoxide dismutase gene provide an animal model of amyotrophic lateral sclerosis. Proc Natl Acad Sci USA 92: 689–693

Robitaille Y, Lopes-Cendes I, Becher M, Rouleau G, Clark AW (1997) The neuropathology of CAG repeat diseases: review and update of genetic and molecular features. Brain Pathol 7: 901–926

Rosen DR, Siddique T, Patterson D, Figlewicz DA, Sapp P, Hentati A, Donaldson D, Goto J, O'Regan JP, Deng H-X, Rahmani Z, Krizus A, McKenna-Yasek D, Cayabyab A, Gaston SM, Berger R, Tanzi RE,

Halperin JJ, Herzfeldt B, Van den Bergh R, Hung W-Y, Bird T, Deng G, Mulder DW, Smyth C, Laing NG, Soriano E, Pericak-Vance MA, Haines J, Rouleau GA, Gusella JS, Horvitz HR, Brown RH Jr (1993) Mutations in Cu/Zn superoxide dismutase gene are associated with familial amyotrophic lateral sclerosis. Nature 362: 59–62

Rosenblum WI (1999) The presence, origin, and significance of Aβ peptide in the cell bodies of neurons. J Neuropath Exp Neurol 58: 575–581

Rothstein JD (1996) Excitotoxicity hypothesis. Neurology 47: S19–S26

Rothstein JD, Dykes-Hoberg M, Corson LB, Becher M, Cleveland DW, Price DL, Culotta VC, Wong PC (1999) The coper chaperone CCS is abundant in neurons and astrocytes in human and rodent brain. J Neurochem 72: 422–429

Rowland LP (1994) Natural history and clinical features of amyotrophic lateral sclerosis and related motor neuron diseases. In: Calne DB (ed) Neurodegenerative diseases. Philadelphia, W. B. Saunders pp. 507–521

Schenk D, Barbour R, Dunn W, Gordon G, Grajeda H, Guido T, Hu K, Huang J, Johnson-Wood K, Khan K, Kholodenko D, Lee M, Liao Z, Lieberburg I, Motter R, Mutter L, Soriano F, Shopp G, Vasquez N, Vandevert C, Walker S, Wogulis M, Yednock T, Games D, Seubert P (1999) Immunization with amyloid-beta attenuates Alzheimer-disease-like pathology in the PDAPP mouse. Nature 400: 173–177

Schilling G, Becher MW, Sharp AH, Jinnah HA, Duan K, Kotzuk JA, Slunt HH, Ratovitski T, Cooper JK, Jenkins NA, Copeland NG, Price DL, Ross CA, Borchelt DR (1999) Intranuclear inclusions and neuritic pathology in transgenic mice expressing a mutant N-terminal fragment of huntingtin. Human Mol Genet 8: 397–407

Schmidt PJ, Rae TD, Pufahl RA, Hamma T, Strain J, O'Halloran TV, Culotta VC (1999) Multiple protein domains contribute to the action of the copper chaperone for superoxide dismutase. J Biol Chem 274: 23719–23725

Selkoe DJ (1999) Translating cell biology into therapeutic advances in alzheimer's disease. Nature 399: A23–A31

Shaw PJ (2000) Biochemical pathology. In: Brown RH Jr, Meininger V, Swash M (eds) Amyotrophic lateral sclerosis. London, Martin Dunitz Ltd, pp. 113–144

Shen J, Bronson RT, Chen DF, Xia W, Selkoe DJ, Tonegawa S (1997) Skeletal and CNS defects in presenilin-1-deficient mice. Cell 89: 629–639

Sherrington R, Rogaev EI, Liang Y, Rogaeva EA, Levesque G, Ikeda M, Chi H, Lin C, Li G, Holman K, Tsuda T, Mar L, Foncin J-F, Bruni AC, Montesi MP, Sorbi S, Rainero I, Pinessi L, Nee L, Chumakov I, Pollen D, Brookes A, Sanseau P, Polinsky RJ, Wasco W, Da Silva HAR, Haines JL, Pericak-Vance MA, Tanzi RE, Roses AD, Fraser PE, Rommens JM, St George-Hyslop PH (1995) Cloning of a gene bearing missense mutations in early-onset familial Alzheimer's disease. Nature 37: 754–760

Simone NL, Bonner RF, Gillespie JW, Emmert-Buck MR, Liotta LA (1998) Laser-capture microdissection: opening the microscopic frontier to molecular analysis. Trends Genet 14: 272–276

Sinha S, Anderson JP, Barbour R, Basl GS, Caccavello R, Davis D, Doan M, Dovey HF, Frigon N, Hong J, Jacobson-Croak K, Jewett N, Kelm P, Knops J, Lieberburg I, Power M, Tan H, Tatsuno G, Tung J, Schenk D, Seubert P, Suomensaari SM, Wang S, Walker D, Zhao J, McConlogue L, John V (1999) Purification and cloning of amyloid precursor protein beta-secretase from human brain. Nature 402: 537–540

Sisodia SS (1992) β-amyloid precursor protein cleavage by a membrane-bound protease. Proc Natl Acad Sci USA 89: 6075–6079

Sisodia SS, Koo EH, Beyreuther K, Unterbeck A, Price DL (1990) Evidence that β-amyloid protein in Alzheimer's disease is not derived by normal processing. Science 248: 492–495

Sisodia SS, Koo EH, Hoffman PN, Perry G, Price DL (1993) Identification and transport of full-length amyloid precursor proteins in rat peripheral nervous system. J Neurosci 13: 3136–3142

St. George-Hyslop PH (1999) Molecular genetics of Alzheimer disease. In: Terry RD, Katzman R, Bick KL, Sisodia SS (eds) Alzheimer disease. Philadelphia, Lippincott Williams & Wilkins, pp. 311–326

St. George-Hyslop PH, Westaway DA (1999) Antibody clears senile plaques. Nature 400: 116–117

Stadtman ER (1992) Protein oxidation and aging. Science 257: 1220–1224

Struchler-Pierrat C, Abramowski D, Duke M, Wiederhold K-H, Mistl C, Rothacher S, Ledermann B, Bürki K, Frey P, Paganetti PA, Waridel C, Calhoun ME, Jucker M, Probst A, Staufenbiel M, Sommer B (1997) Two amyloid precursor protein transgenic mouse models with Alzheimer disease-like pathology. Proc Natl Acad Sci USA 94: 13287–13292

Suzuki N, Cheung TT, Cai X-D, Odaka A, Otvos L Jr, Eckman C, Golde TE, Younkin SG (1994) An increased percentage of long amyloid β protein secreted by familial amyloid β protein precursor (βAPP_{717}) mutants. Science 264: 1336–1340

Sze C-I, Troncoso JC, Kawas CH, Mouton PR, Price DL, Martin LJ (1997) Loss of the presynaptic vesicle protein synaptophysin in hippocampus correlates with cognitive decline in Alzheimer's disease. J Neuropathol Exp Neurol 56: 933–944

Thinakaran G, Borchelt DR, Lee MK, Slunt HH, Spitzer L, Kim G, Ratovitski T, Davenport F, Nordstedt C, Seeger M, Hardy J, Levey AI, Gandy SE, Jenkins N, Copeland N, Price DL, Sisodia SS (1996) Endoproteolysis of presenilin 1 and accumulation of processed derivatives in vivo. Neuron 17: 181–190

Tremblay P, Meiner Z, Galou M, Heinrich C, Petromilli C, Lisse T, Cauetano J, Torchia M, Mobley W, Bujard H, DeArmond SJ, Prusiner SB (2000) Doxycycline control of prion protein transgene expression modulates prion disease in mice. Proc Natl Acad Sci USA 95: 12580–12585

Tu P-H, Raju P, Robinson KA, Gurney ME, Trojanowski JQ, Lee VMY (1996) Transgenic mice carrying a human mutant superoxide dismutase transgene develop neuronal cytoskeletal pathology resembling human amyotrophic lateral sclerosis lesions. Proc Natl Acad Sci USA 93: 3155–3160

Valentine JS, Gralla EB (1997) Delivering copper inside yeast and human cells. Science 278: 817–818

Vassar R, Bennett BD, Babu-Khan S, Kahn S, Mendiaz EA, Denis P, Teplow DB, Ross S, Amarante P, Loeloff R, Luo L, Fisher S, Fuller J, Edenson S, Lile J, Jarosinski MA, Biere AL, Curran E, Burgess T, Louis J-C, Collins F, Treanor J, Rogers G, Citron M (1999) β-secretase cleavage of alzheimer's amyloid precursor protein by the transmembrane aspartic protease BACE. Science 286: 735–741

Waggoner DJ, Bartnikas TB, Gitlin JD (1999) The role of copper in neurodegenerative disease. Neurobiol Dis 6: 221–230

Walker LC, Cork LC (1999) The neurobiology of aging in nonhuman primates. In: Terry RD, Katzman R, Bick KL, Sisodia SS (eds) Alzheimer disease. Philadelphia, Lippincott Williams & Wilkins pp. 233–244

Weidemann A, König G, Bunke D, Fischer P, Salbaum JM, Masters CL, Beyreuther K (1989) Identification, biogenesis, and localization of precursors of Alzheimer's disease A4 amyloid protein. Cell 57: 115–126

Whitehouse PJ, Price DL, Struble RG, Clark AW, Coyle JT, DeLong MR (1982) Alzheimer's disease and senile dementia: loss of neurons in the basal forebrain. Science 215: 1237–1239

Wiedau-Pazos M, Goto JJ, Rabizadeh S, Gralla EB, Roe JA, Lee MK, Valentine JS, Bredesen DE (1996) Altered reactivity of superoxide dismutase in familial amyotrophic lateral sclerosis. Science 271: 515–518

Williamson TL, Bruijn LI, Zhu O, Anderson KL, Anderson SD, Julien J-P, Cleveland DW (1998) Absence of neurofilaments reduces the selective vulnerability of motor neurons and slows disease caused by a familial amyotrophic lateral sclerosis-linked superoxide dismutase 1 mutant. Proc Natl Acad Sci USA 95: 9631–9636

Williamson TL, Cleveland DW (1999) Slowing of axonal transport is a very early event in the toxicity of ALS-linked SOD1 mutants to motor neurons. Nature Neurosci 2: 50–56

Wilson CA, Doms RW, Lee VMY (1999) Intracellular APP processing and A-beta production in Alzheimer disease. J Neuropathol Exp Neurol 58: 787–794

Wolfe MS, Xia W, Ostazewski BL, Diehl TS, Kimberly WT, Selkoe DJ (1999) Two transmembrane aspartates in presenilin-1 required for presenilin endoproteolysis and γ-secretase activity. Nature 398: 513–517

Wong PC, Borchelt DR (1995) Motor neuron disease caused by mutations in superoxide dismutase 1. Curr Opin Neurol 8: 294–301

Wong PC, Pardo CA, Borchelt DR, Lee MK, Copeland NG, Jenkins NA, Sisodia SS, Cleveland DW, Price DL (1995) An adverse property of a familial ALS-linked SOD1 mutation causes motor neuron disease characterized by vacuolar degeneration of mitochondria. Neuron 14: 1105–1116

Wong PC, Rothstein JD, Price DL (1998) The genetic and molecular mechanisms of motor neuron disease. Curr Opin Neurobiol 8: 791–799

Wong PC, Waggoner D, Subramaniam J, Tessarollo L, Bartnikas TB, Culotta VC, Price DL, Rothstein J, Gitlin JD (2000) Copper chaperone for superoxide dismutase is essential to activate mammalian Cu/Zn superoxide dismutase. Proc Natl Acad Sci USA 97: 2886–2891

Wong PC, Zheng H, Chen H, Becher MW, Sirinathsinghji DJS, Trumbauer ME, Chen HY, Price DL, Van der Ploeg LHT, Sisodia SS (1997) Presenilin 1 is required for *Notch1* and *Dll1* expression in the paraxial mesoderm. Nature 387: 288–292

Yan R, Blenkowski MJ, Shuck ME, Mlao H, Tory MC, Pauley AM, Brashler JR, Stratman NC, Mathews WR, Buhl AE, Carter DB, Tomasselli AG, Parodl LA, Heinrikson RL, Gurney ME (1999) Membrane-anchored aspartyl protease with Alzheimer's disease beta-secretase activity. Nature 402: 533–537

Yim MB, Kang JH, Yim HS, Kwak HS, Chock PB, Stadtman ER (1996) A gain-of-function of an amyotrophic lateral sclerosis-associated Cu,Zn-superoxide dismutase mutant: An enhancement of free radical formation due to a decrease in K_m for hydrogen peroxide. Proc Natl Acad Sci USA 93: 5709–5714

Zhang B, Tu P-H, Abtahian F, Trojanowski JQ, Lee VMY (1997) Neurofilaments and orthograde transport are reduced in ventral root axons of transgenic mice that express human SOD1 with a G93A mutation. J Cell Biol 139: 1307–1315

Zheng H, Jiang M-H, Trumbauer ME, Sirinathsinghji DJS, Hopkins R, Smith DW, Heavens RP, Dawson GR, Boyce S, Conner MW, Stevens KA, Slunt HH, Sisodia SS, Chen HY, Van der Ploeg LHT (1995) β-amyloid precursor protein-deficient mice show reactive gliosis and decreased locomotor activity. Cell 81: 525–531

Zoghbi HY, Orr HT (1999) Polyglutamine diseases: protein cleavage and aggregation. Curr Opin Neurobiol 9: 566–570

Pathogenesis and Mechanism
of Cerebral Amyloidosis in APP Transgenic Mice

M. Jucker, M. Calhoun, A. Phinney, M. Stalder, L. Bondolfi, D. Winkler,
M. Herzig, M. Pfeifer, S. Boncristiano, M. Tolnay, A. Probst, T. Deller,
D. Abramowski, K.-H. Wiederhold, C. Sturchler-Pierrat, B. Sommer,
and M. Staufenbiel

Summary

Transgenic mice that overexpress mutant human amyloid precursor protein (APP)
exhibit one hallmark of Alzheimer's disease pathology, namely the extracellular
deposition of amyloid in plaques and vessels. In an effort to study the impact of
cerebral amyloidosis on neurodegeneration, we have shown that amyloid plaque
formation in APP transgenic mice is accompanied by region-specific neuron loss,
synaptic changes, alterations in the cholinergic system, and a severe disruption of
neuronal circuits. The deposition of amyloid in the vessel wall leads to smooth
muscle cell degeneration and spontaneous hemorrhagic stroke. In humans several
mechanisms may contribute to cerebral amyloidosis. Results from transgenic
mice, however, suggest that a neuronal source of APP/Aβ is sufficient for the
development of both amyloid plaques and cerebrovascular amyloid. Moreover,
our results implicate neuronal transport and drainage mechanisms rather than
local production or blood uptake of Aβ as a primary mechanism underlying cerebral amyloidosis in these mice. Aging and APP expression levels have been suggested to be key factors that potentiate amyloid deposition. But there are several
other risk factors, such as apolipoprotein E and transforming growth factor
TGFβ1, that have been identified and analyzed in association with APP transgenic
mice. In conclusion, transgenic mouse models of cerebral amyloidosis have provided many clues about the significance and mechanism of cerebral amyloidosis.
The continued analysis of these mice will provide the tools to develop therapeutic
intervention in AD, cerebral amyloid angiopathy, and hemorrhagic stroke.

Introduction

Cerebral amyloidosis and neurofibrillary tangles are hallmark pathologies of Alzheimer's disease (AD). The principal constituent of amyloid deposits is the
amyloid-β (Aβ) peptide. Aβ is a 40 or 42 amino acid-long peptide that is derived
from the amyloid-β precursor protein (APP; Kang et al. 1987; for review, Price
and Sisodia 1998; Selkoe 1998). Neurofibrillary tangles are primarily composed
of hyperphosphorylated tau (Spillantini and Goedert 1998).

Although the majority of AD cases are of sporadic nature, there are families
in which the disease segregates in an autosomal dominant fashion that leads to

Research and Perspectives in Alzheimer's Diseases
Beyreuther/Christen/Masters (Eds.)
Neurodegenerative Disorders
© Springer-Verlag Berlin Heidelberg 2001

an early onset of the disease. The first genetic loci for familial AD were identified as missense mutations in APP. Thereafter, several other mutations in APP, as well as mutations in presenilin 1 and presenilin 2, which underlie familial AD, have been identified (Price and Sisodia 1998). It has recently become clear that all pathogenic APP and presenilin 1 and 2 mutations alter the processing of APP either through increased production of total $A\beta$ or specifically of the longer $A\beta_{42}$ isoform (Selkoe 1997; Hardy 1997).

These findings clearly indicate that amyloidogenic APP processing is crucial to the pathophysiology of AD. In addition to plaques and tangles, the AD brain is characterized by extensive neural degeneration. However, the relationship between amyloidogenic APP processing and AD neurodegeneration remains unclear. Both the neocortex and hippocampus of AD patients undergo marked atrophy, region-specific neuron loss, synapse loss, and a striking reduction in cholinergic markers. Furthermore, these deficits have all been shown to be related to the cognitive status of individual patients, with synaptic alterations being the best morphological correlate to AD dementia (Bartus et al. 1982; Terry et al. 1991; West et al. 1994; DeKosky et al. 1996; Gomez-Isla et al. 1997; Mouton et al. 1998). Yet, despite intensive interest in a potential association between the amyloid plaques, vascular amyloid, tangles and neurodegeneration observed in AD brain, a clear relationship has remained elusive.

Although these questions are very important to address for the development of therapeutic interventions, until recently progress in this area has been slow due to the lack of suitable animal models. Past studies have largely been based upon the analysis of post-mortem AD brain tissue, which often reflects the end-stage of the disease. The recent progress in the development of transgenic mouse models has come largely from advanced understanding of the molecular genetics of the disease. Several groups generated transgenic mice that exhibit age-related deposition of cerebral amyloid similar to AD through expression of mutated human APP (Games et al. 1995; Hsiao et al. 1996; Sturchler-Pierrat et al. 1997). These mice have learning impairments and deficits in synaptic transmission and/or long term potentiation (Chapman et al. 1999; Hsia et al. 1999). However, the mechanisms of amyloid deposition and how the formation of amyloid deposits relates to AD neurodegeneration are still not clear. These mouse models now provide a new focus for many rigorous studies concerning the significance and mechanisms of cerebral amyloidoisis.

Impact of Cerebral Amyloid on Neurodegeneration

Neuron Loss

To study the impact of amyloid plaque formation on neurodegeneration, we and others have used modern stereological techniques to relate neuron number to amyloid burden in APP transgenic mice (Irizarry et al. 1997a, b; Calhoun et al. 1998). In APP23 mice, we have shown that amyloid plaque formation is accompanied by a region-specific neuron loss. While 18-month-old APP23 mice with a high plaque load reveal a 25 % loss of CA1 hippocampal neurons, no significant neuron loss in neocortex could be detected when compared to age-matched controls (Calhoun et al. 1998). Initially, CA1 neuron loss in APP23 mice appears to disagree with the reported non-significant loss of CA1 neurons in PrP-APP mice (Irizarry et al. 1997a). However, a closer look actually shows agreement between the two studies because only a 4 % plaque load was reported in the PrP-APP mice versus a 10–25 % plaque load in APP23 mice. Yet in another mouse with an amyloid burden similar to APP23 mice, no loss of CA1 hippocampal neurons was reported (Irizarry et al. 1997b). However, PD-APP mice have significant amounts of diffuse amyloid (Masliah et al. 1996; Irizarry et al. 1997b), which is considered to be the non-toxic form of amyloid deposition, whereas 90 % of the amyloid in APP23 mice is of the more disruptive, congophilic nature (Calhoun et al. 1998; Stalder et al. 1999).

We have recently reinvestigated the apparent lack of neuron loss in neocortex of APP23 mice using a very old cohort of mice (Bondolfi et al. unpublished results). Again, we found no difference in neocortical neuron number between 27-month-old APP23 mice and agematched control mice, although there was a correlation between neuron number and amyloid plaque load in the APP23 mice. This puzzling observation was resolved when we found a significant increase in neocortical neuron number in 8-month-old APP23 mice compared to 8-month-old control mice. These results in younger mice suggest that APP23 mice have more neocortical neurons compared to non-transgenic controls but lose neurons in the process of amyloid plaque formation. This interpretation is consistent with a neurotrophic/neuroprotective function of APP and a neurotoxic function of Aβ (Mucke et al. 1996; Yankner 1996).

Synaptic Changes and Aberrant Synaptogenesis

A focal loss of synapses and the presence of large dystrophic synaptic boutons around congophilic plaques have consistently been reported in several APP transgenic mice (Games et al. 1995; Masliah et al. 1996; Phinney et al. 1999). Yet, inconsistent results have been published within and among transgenic mouse lines regarding total synapse numbers using quantitative synaptophysin immunostaining. While some studies reported a loss of synaptic boutons in hippocampal subregions (Games et al. 1995, Hsia et al. 1999), others have reported no change (Irizarry et al. 1999a, b). We and others have used modern stereology to determine the total number of synaptophysin-positive and ChAT-positive synapses in neocortex of APP23 and PrP-APP transgenic mice, respectively (Calhoun et al., unpublished results; Wong et al. 1999). In both studies mice with only a light amyloid burden were analyzed, and in both studies an increase in synaptic bouton number was found. These results confirm previous observations of a synaptotrophic effect of APP overexpression in transgenic mice (Mucke et al. 1994). Whether the heavy deposition of amyloid in older mice leads to a secondary reduction in synapse number is currently under investigation. If found to be true, this finding would be very consistent with the observation of neuron numbers in these mice.

We have also used anterograde tracing techniques to study the effect of amyloid deposition on synapses and neural connectivity. For this purpose we studied the axonal projection from the entorhinal cortex to the hippocampus in APP23 mice (Phinney et al. 1999). We found that entorhinal axons form dystrophic boutons around dense core amyloid plaques in the entorhinal termination zone of the hippocampus. Moreover, entorhinal boutons were associated with amyloid in ectopic locations within the hippocampus, the thalamus, and white matter tracts. Many of these ectopic entorhinal boutons were immunopositive for the growth-associated-protein GAP-43 and showed light and electron microscopic characteristics of synaptic terminals (Phinney et al. 1999). These findings demonstrate that cerebral amyloid deposition leads to aberrant sprouting and the formation of ectopic synaptic terminals. Although sprouting has been observed in AD, our results suggest that the magnitude and aberrant nature of this sprouting may not have been appreciated in human. Such inappropriate sprouting may severely disrupt normal neuronal circuitry, which in turn may contribute to AD dementia.

Alterations in the Cholinergic System

Dystrophic cholinergic fibers have been observed in the vicinity of amyloid plaques in APP transgenic mice (Sturchler-Pierrat et al. 1997; Wong et al. 1999). No studies of cholinergic enzyme activity has yet been reported, although acetylcholinesterase fiber staining and immunostaining for choline acetyltransferase clearly show a disruption of the cholinergic fiber network in the neocortex of APP transgenic mice with amyloid plaques (Wong et al. 1999; Boncristiano et al., in preparation). Interestingly, no loss of basal cholinergic forebrain neurons has been reported in PrP-APP and APP23 mice, despite significant amyloid deposition in the neocortex and hippocampus (Westerman et al. 1999; Boncristiano et al., in preparation).

Cerebral Amyloid Angiopathy and Cerebral Hemorrhage

In addition to amyloid plaques, APP transgenic mice accumulate amyloid in the vasculature (cerebral amyloid angiopathy, CAA). While significant amounts of CAA have been reported in APP23 mice (Calhoun et al. 1999), CAA does not appear to be prominent in other APP transgenic mice (Su and Ni 1998; Kane et al. 2000). The reasons for this discrepancy are not entirely clear, yet differences in APP expression levels and genetic backgrounds are plausible explanations.

CAA in APP23 mice is very similar to that found in humans (Calhoun et al. 1999). All types of vessels are affected, although amyloid deposition occurs preferentially in arterioles and capillaries. Within individuals vessels amyloid deposition shows a wide heterogeneity, ranging from deposits confined to the vessel basement membrane to large plaque-like extrusions into the neuropil. The progressive deposition of amyloid in the vessel wall leads to the loss of smooth muscle cells, the formation of microaneurism, and multiple microhemorrhages. We have recently collected evidence for recurrent bleedings and the occurrence of large hemorrhagic strokes in aged APP23 mice (Winkler et al. 2000). Interestingly, before CAA leads to blood vessel rupture and hemorrhagic stroke, neuron loss, synaptic abnormalities, and activation of microglia are observed in the vicinity of amyloid-coated vessels (Calhoun et al. 1999). This finding shows that vascular amyloid can lead to neurodegeneration, either through a toxic effect of the amyloid or through a diminished transport of nutrients to the neuropil. Moreover, cerebrovascular amyloid, which is very similar to amyloid plaques, can induce aberrant sprouting (Phinney et al. 1999) and thus contribute to the disruption of neural connectivity.

Mechanism of Cerebral Amyloid Deposition

Cerebral amyloidosis, especially CAA, in APP23 mice is of particular interest considering the strict neuronal source of APP, which was achieved by using a neuron-specific promoter (Sturchler-Pierrat et al. 1997; Calhoun et al. 1999). Thus, in mice a neuronal source of APP appears sufficient to induce not only extracellular amyloid plaques but also CAA. This conclusion is further substantiated by the finding that APP23 mice on an *App*-null background develop a similar degree of amyloid plaques and CAA (Calhoun et al. 1999). The neuronal source of APP/Aβ in APP23 mice, together with the observation of substantial amyloid deposits in brain regions without detectable transgene expression (Calhoun et al. 1999), suggests that axonal transport and/or circulation of Aβ play a role in cerebral amyloidosis.

It has previously been shown that APP undergoes axonal transport and that Aβ-containing APP fragments accumulate at synaptic sites (Buxbaum et al. 1998). Thus, axonal transport and synaptic release of Aβ may explain the consistent observation of amyloid deposits in the molecular layer of the dentate gyrus (e. g., Su and Ni 1998; Irizarry et al. 1997a, b), which is the termination zone of entorhinal fibers in mice (Stanfield et al. 1979; Deller et al. 1999). Similarly, corticothalamic axonal transport could then explain the significant amyloid deposition in the thalamus of APP23 mice, an area with no detectable transgene expression (Calhoun et al. 1999).

Further clues regarding the mechanism of cerebral amyloidosis come from the analysis of human Aβ levels in cerebrospinal fluid (CSF) and blood plasma of APP23 mice (Calhoun et al. 1999). Results revealed about 10 times higher human Aβ levels in the CSF of APP23 mice compared to Aβ levels in CSF of normal human and AD patients. In contrast, only trace amounts of human Aβ were found in the blood of APP23 mice. This observation is interesting in light of two previously established transgenic mouse lines that show exceptionally high levels of Aβ in the plasma but do not develop amyloid plaques and CAA (Fukuchi et al.

1996; Kawarabayashi et al. 1996). Thus, the flow of Aβ from neurons to the CSF and/or drainage of Aβ from the CSF must be considered as factors in the mechanism of cerebral amyloid deposition. Interestingly, it has been suggested that Aβ in the interstitial fluid (ISF) and CSF drains along periarterial spaces into cervical lymph nodes (Weller et al. 1998). Along this drainage pathway, Aβ may accumulate and form cerebrovascular amyloid.

Amyloid deposition in APP transgenic mice is clearly dependent upon APP and Aβ expression levels (Hsiao et al. 1996; Sturchler-Pierrat et al. 1997; Calhoun et al. 1999). This finding is in line with the observation that APP/PS1 double transgenic mice develop cerebral amyloidosis well before APP transgenic mice (Borchelt et al. 1997; Holcomb et al. 1998). APP transgenic mice have also been crossed with mice deficient in apolipoprotein E (ApoE) and mice overexpressing transforming growth factor TGFβ1. In contrast to the previously mentioned APP/PS1 double transgenic mice, APP transgenic mice on a ApoE-null background show no change in Aβ production. Nevertheless, these mice show a remarkable decrease of congophilic amyloid, suggesting that ApoE promotes the deposition and/or fibrillization of Aβ (Bales et al. 1997, 1999; Holtzmann et al. 2000). Similarly, no change in Aβ production is observed in crosses between APP transgenic mice and TGFβ1 transgenic mice. However, these double transgenic mice show an increase in CAA that may be caused by the effect of TGFβ1 on the production of extracellular matrix molecules, which in turn may have an effect on the deposition and/or clearance of Aβ (Wyss-Coray et al. 1997).

Finally, a role for microglia in cerebral amyloidosis, either in amyloid deposition or clearance, has been suggested. A tight association of microglia with congophilic amyloid and a lack of microglia activation around diffuse plaques have been described in APP transgenic mice (Frautschy et al. 1998; Stalder et al. 1999). Although membrane-bound bundles of amyloid fibrils surrounded by the microglial cytoplasm have been observed, recent studies have shown that such amyloid is in fact extracellular (Stalder et al. 2000). Moreover, no evidence for amyloid phagocytosis by microglia has been found. Consistently, crossing of APP transgenic mice with mice deficient in the class A scavenger receptor (a receptor that has been shown to be involved in amyloid phagocytosis in cell culture) does not affect amyloid deposition in vivo (Huang et al. 1999). Finally, and perhaps most interestingly, a recent study reported that immunization with Aβ prevented the deposition of amyloid in PD-APP transgenic mice, and similar immunization in older mice reduced the extent of existing amyloid deposits greatly (Schenk et al. 1999). Strikingly, a close association of activated MHC class II-positive microglia and amyloid deposits was observed in these immunized mice. This finding raises the possibility that microglia may be able to phagocytose and degrade IgG-tagged amyloid fibrils.

Conclusion

Using a transgenic approach, we have shown that cerebral amyloidosis in APP transgenic mice is sufficient to induce several types of neurodegeneration observed in AD brain. Although neurodegeneration in transgenic mice is quantitatively less severe than in AD brain, the data from these mice support the hypothesis that amyloidogenic APP processing with amyloid deposition plays a primary role in AD pathogenesis. In terms of mechanisms, results from transgenic mice suggest that a neuronal source of Aβ is sufficient for the deposition of Aβ in plaques and vasculature and that transport and clearance of Aβ play an important role. Although more comprehensive studies are necessary, the present results have already improved the current understanding of the pathophysiology of AD and CAA and provide a basis for the development of effective interventions.

Acknowledgments

This work was supported by grants from the Swiss National Foundation; the Horten Foundation, Madonna del Piano, Switzerland; the Roche Foundation, Basel; Switzerland; the Fritz Thyssen Foundation, Cologne, Germany; the VerUm Foundation, Munich, Germany; the DFG (SFB 505), Germany; and the AETAS-Foundation for Research into Ageing (with the financial support of the Loterie Romande), Geneva, Switzerland.

References

Bales KR, Verina T, Dodel RC, Du Y, Altstiel L, Bender M, Hyslop P, Johnstone EM, Little SP, Cummins DJ, Piccardo P, Ghetti B, Paul SM (1997) Lack of apolipoprotein E dramatically reduces amyloid β-peptide deposition. Nature Genet 17: 263–264

Bales KR, Verina T, Cummins DJ, Du Y, Dodel RC, Saura J, Fishman CE, DeLong CA, Piccardo P, Petegniel V, Ghetti B, Paul SM (1999) Apolipoprotein E is essential for amyloid deposition in the APP^{V717F} transgenic mouse model of Alzheimer's disease. Proc Natl Acad Sci USA 96: 15233–15238

Bartus RT, Dean RL, Beer B, Lippa AS (1982) The cholinergic hypothesis of geriatric memory dysfunction. Science 217: 408–414

Borchelt DR, Ratovitski T, van Lare J, Lee MK, Gonzales V, Jenkins NA, Copeland NG, Price DL, Sisodia SS (1997) Accelerated amyloid deposition in the brains of transgenic mice coexpressing mutant presenilin 1 and amyloid precursor proteins. Neuron 19: 939–45

Buxbaum JD, Thinakaran G, Koliatsos V, O'Callahan J, Slunt HH, Price DL, Sisodia SS (1998) Alzheimer amyloid protein precursor in the rat hippocampus: transport and processing through the perforant path. J Neurosci 18: 9629–9637

Calhoun M, Wiederhold K, Abramowski D, Phinney A, Probst A, Sturchler-Pierrat C, Staufenbiel M, Sommer B, Jucker M (1998) Neuron loss in APP transgenic mice. Nature 395: 766–756

Calhoun ME, Burgermeister P, Phinney AL, Stalder M, Tolnay M, Wiederhold KH, Abramowski D, Stürchler-Pierrat C, Sommer B, Staufenbiel M, Jucker M (1999) Neuronal overexpression of mutant amyloid precursor protein results in prominent deposition of cerebrovascular amyloid. Proc Natl Acad Sci USA 96: 14088–14093

Chapman PF, White GL, Jones MW, Cooper-Blacketer D, Marshall VJ, Irizarry M, Younkin L, Good MA, Bliss TV, Hyman BT, Younkin SG, Hsiao KK (1999) Impaired synaptic plasticity and learning in aged amyloid precursor protein transgenic mice. Nature Neurosci 2: 271–276

DeKosky ST, Scheff SW, Styren SD (1996). Structural correlates of cognition in dementia: quantification and assessment of synapse change. Neurodegeneration 5: 417–421

Deller T, Drakew A, Frotscher M (1999) Different primary target cells are important for fiber lamination in the fascia dentata: a lesson from reeler mutant mice. Exp Neurol 156: 239–253

Frautschy SA, Yang F, Irrizarry M; Hyman B, Saido TC, Hsiao K, Cole GM (1998) Microglial response to amyloid plaques in APPsw transgenic mice. Am J Pathol 152: 307–317

Fukuchi K, Ho L, Younkin SG, Kunkel DD, Ogburn CE, LeBoeuf RC, Furlong CE, Deeb SS, Nochlin D, Wegiel J, Wisniewski HM, Martin GM (1996) High levels of circulating β-amyloid peptide do not cause cerebral β-amyloidosis in transgenic mice. Am J Pathol 149: 219–227

Games D, Adams D, Alessandrini R, Barbour R, Berthelette P, Blackwell C, Carr T, Clemens J, Donaldson T, Gillespie F, Guido T, Hagopian S, Johnson-Wood K, Khan K, Lee M, Leibowitz P, Lieberburg I, Little S, Masliah E, McConlogue L, Montoya-Zavala M, Mucke L, Paganini L, Penniman E, Power M, Schenk D, Seubert P, Synder B, Soriano F, Tan H, Vitale J, Wadsworth S, Wolozin B, Zhao J (1995) Alzheimer-type neuropathology in transgenic mice overexpressing V717F β-amyloid precursor protein. Nature 373: 523–527

Gomez Isla T, Hollister R, West H, Mui S, Growdon JH, Petersen RC, Parisi JE, Hyman BT (1997) Neuronal loss correlates with but exceeds neurofibrillary tangles in Alzheimer's disease. Ann Neurol 41: 17–24

Hardy J (1997) Amyloid, the presenilins and Alzheimer's disease. Trends Neurosci 20: 154–159

Holcomb L, Gordon MN, McGowan E, Yu X, Benkovic S, Jantzen P, Wright K, Saad I, Mueller R, Morgan D, Sanders S, Zehr C, O'Campo K, Hardy J, Prada CM, Eckman C, Younkin S, Hsiao K, Duff K (1998). Accelerated Alzheimer-type phenotype in transgenic mice carrying both mutant amyloid precursor protein and presenilin 1 transgenes. Nature Med 4: 97–100

Holtzman DM, Bakes KB, Tenkova T, Fagan AM, Parsadanian M, Sartorius LJ, Mackey B, Olney J, McKeel D, Wozniak D, Paul SM (2000) Apolipoprotein E isoform-dependent amyloid deposition and neuritic degeneration in a mouse model of Alzheimer's disease. Proc Natl Acad Sci USA 97: 2892–2897

Hsia AY, Masliah E, McConlogue L, Yu GQ, Tatsuno G, Hu K, Kholodenko D, Malenka RC, Nicoll RA, Mucke L (1999) Plaque-independent disruption of neural circuits in Alzheimer's disease mouse models. Proc Natl Acad Sci USA 96: 3228–3233

Hsiao K, Chapman P, Nilsen S, Eckman C, Harigaya Y, Younkin S, Yang F, Cole G (1996) Correlative memory deficits, Aβ elevation, and amyloid plaques in transgenic mice. Science 274: 99–102

Huang F, Buttini M, Wyss-Coray T, McConlogue L, Kodma T, Pitas RE, Mucke L (1999) Elimination of the class A scavenger receptor does not affect amyloid plaque formation or neurodegeneration in transgenic mice expressing human amyloid precursor protein. Am J Pathol 155: 1741–1747

Irizarry MC, McNamara M, Fedorchak K, Hsiao K, Hyman BT (1997a) APP$_{Sw}$ transgenic mice develop age-related Aβ deposits and neuropil abnormalities, but no neuronal loss in CA1. J Neuropathol Exp Neurol 56: 965–973

Irizarry MC, Soriano F, McNamara M, Page KJ, Schenk D, Games D, Hyman BT (1997b) Aβ deposition is associated with neuropil changes, but not with overt neuronal loss in the human amyloid precursor protein V717F (PDAPP) transgenic mouse. J Neurosci 17: 7053–7059

Kane MD, Lipinski WJ, Callahan MJ, Bian F, Durham RA, Schwarz RD, Roher AE, Walker LC (2000) Evidence for seeding of β-Amyloid by intracerebral infusion of Alzheimer brain extracts in βAPP-transgenic mice. J Neurosci 20: 3606–3611

Kang J, Lemaire HG, Unterbeck A, Salbaum JM, Masters CM, Grzeschik K-H, Multhaup G, Beyreuther K, Müller Hill B (1987) The precursor of Alzheimer's disease amyloid A4 protein resembles a cell-surface receptor. Nature 325: 733–736

Kawarabayashi T, Shoji M, Sato M, Sasaki A, Ho L, Eckman CB, Prada CM, Younkin SG, Kobayashi T, Tada N, Matsurbara E, Iizuka T, Harigaya T, Kasai K (1996) Accumulation of β-amyloid fibrils in pancreas of transgenic mice. Neurobiol Aging 17: 215–222

Masliah E, Sisk A, Mallory M, Mucke L, Schenk D, Games D (1996) Comparison of neurodegenerative pathology in transgenic mice overexpressing V717F β-amyloid precursor protein and Alzheimer's disease. J Neurosci 16: 5795–5811

Mouton PR, Martin LJ, Calhoun ME, Dal Forno G, Price D (1998) Cognitive decline strongly correlates with cortical atrophy in Alzheimer's dementia. Neurobiol Aging 19: 371–377

Mucke L, Masliah E, Johnson WB, Ruppe MD, Alford M, Rockenstein EM, Forss-Petter S, Pietropaolo M, Mallory M, Abraham CR (1994) Synaptotrophic effects of human amyloid β protein precursors in the cortex of transgenic mice. Brain Res 666: 151–167

Mucke L, Abraham CR, Masliah E (1996) Neurotrophic and neuroprotective effects of hAPP in transgenic mice. Ann NY Acad Sci 777: 82–88

Phinney AL, Deller T, Stalder M, Calhoun ME, Frotscher M, Sommer B, Staufenbiel M, Jucker M (1999) Cerebral amyloid induces aberrant axonal sprouting and ectopic terminal formation in amyloid precursor protein transgenic mice. J Neurosci 19: 8552–8559

Price DL, Sisodia SS (1998) Mutant genes in familial Alzheimer's disease and trangenic models. Ann Rev Neurosci 21: 479–505

Schenk D, Barbour R, Dunn W, Gordon G, Grajeda H, Guido T, Hu K, Huang J, Johnson-Wood K, Khan K, Kholodenko D, Lee M, Liao Z, Lieberburg I, Motter R, Mutter L, Soriano F, Shopp G, Vasquez N, Vandervert C, Walker S, Wogulis M, Yednock T, Games D, Seubert P (1999) Immunization with amyloid-β attenuates Alzheimer-disease-like pathology in the PDAPP mouse. Nature 400: 173–177

Selkoe DJ (1997) Alzheimer's disease: genotypes, phenotypes, and treatments. Science 275: 630–631

Selkoe DJ (1998) The cell biology of beta-amyloid precursor protein and presenilin in Alzheimer's disease. Trends Cell Biol 8: 447–453

Spillantini MG, Goedert M (1998) Tau protein pathology in neurodegenerative diseases. Trends Neurosci 21: 428–432

Stalder M, Phinney A, Probst A, Sommer B, Staufenbiel M, Jucker M (1999) Association of microglia with amyloid plaques in brains of APP23 transgenic mice. Am J Pathol 154: 1673–1684

Stalder M, Seller T, Staufenbiel M, Jucker M (2000) 3D-Reconstruction of microglia and amyloid in APP23 transgenic mice: No evidence of intracellular amyloid. Neurobiol. Aging, in press

Stanfield BB, Caviness VSJ, Cowan WM (1979) The organization of certain afferents to the hippocampus and dentate gyrus in normal and reeler mice. J Comp Neurol 185: 461–483

Sturchler-Pierrat C, Abramowski D, Duke M, Wiederhold KH, Mistl C, Rothacher S, Ledermann B, Burki K, Frey P, Paganetti PA, Waridel C, Calhoun ME, Jucker M, Probst A, Staubenbiel M, Sommer B (1997) Two amyloid precursor protein transgenic mouse models with Alzheimer disease-like pathology. Proc Natl Acad Sci USA 94: 13287–13292

Su Y, Ni B (1998) Selective deposition of amyloid-β protein in the entorhinal-dentate projection of a transgenic mouse model of Alzheimer's disease. J Neurosci Res 53: 177–186

Terry RD, Masliah E, Salmon DP, Butters N, DeTeresa R, Hill R, Hansen LA, Katzman R (1991) Physical basis of cognitive alterations in Alzheimer's disease: synapse loss is the major correlate of cognitive impairment. Ann Neurol 30: 572–580

Weller RO, Massey A, Newman TA, Hutchings M, Kuo Y-M, Roher AE (1998) Amyloid β accumulates in putative interstitial fluid drainage pathways in Alzheimer's disease. Am J Pathol 153, 725–33

West MJ, Coleman PD, Flood DG, Troncoso JC (1994) Differences in the pattern of hippocampal neuronal loss in normal ageing and Alzheimer's disease. Lancet 344: 769–772

Westerman MA, Cooper-Blacketer D, Hsiao KK, Low WC (1999) The cholinergic system in a transgenic mouse model of Alzheimer's disease. Soc Neurosci Abstr. 25: 2117

Winkler DT, Bondolfi L, Herzig MC, Jann L, Calhoun ME, Wiederhold K-H, Tolnay M, Staufenbiel M, Jucker M (2000) Spontaneous hemorrhagic stroke in a mouse model of cerebral amyloid angiopathy. J. Neurosci, in press

Wong TP, Debeir T, Duff K, Cuello AC (1999) Reorganization of cholinergic terminals in the cerebral cortex and hippocampus in transgenic mice carrying mutated presenilin-1 and amyloid precursor protein transgenes. J Neurosci 19: 2706–2716

Wyss-Coray T, Masliah E, Mallory M, McConlogue L, Johnson-Wood K, Lin C, Mucke (1997) Amyloidogenic role of cytokine TGF-β1 in transgenic mice and in Alzheimer's disease. Nature 389: 603–606

Yankner B (1996) Mechanisms of neuronal degeneration in Alzheimer's disease. Neuron 16: 921–932

Alzheimer's Disease: Physiological and Pathogenetic Role of the Amyloid Precursor Protein (APP), its Aβ-Amyloid Domain and Free Aβ-Amyloid Peptide

K. Beyreuther and C. L. Masters

Summary

To understand synaptic loss and neurodegeneration in Alzheimer's disease, we have tried to consider the physiological functions of the amyloid precursor protein (APP), its Aβ-amyloid domain and of free Aβ peptide. The latter is a normal metabolic product of APP and the principal subunit of the amyloid plaques that are characteristic of Alzheimer's disease. From studies in transgenic *Drosophila melanogaster* and primary mammalian neurons, we suggest that, in neurons, APP exhibits as a physiological function the negative regulation of synaptic strength whereas in nonneuronal cells APP appears to regulate cell-cell and cell-matrix adhesion.

Since the axonal transport of APP is dependent on the Aβ domain, this finding suggests that the Aβ sequence could function as an axonal sorting signal of APP. It also indicates that the Aβ region could bind to molecules that control the recruitment of APP into axonally transported vesicles.

In neurons, metabolism of APP releasing the Aβ peptide was found to occur at all sorting stations, such as at the ER/cisGolgi and TGN/endosomes producing intracellular Aβ peptide as well as at the cell surface leading to secretory Aβ peptide. Regarding the Aβ species generated in the different neuronal compartments, the long form of Aβ (Aβ42) is produced in the ER/cisGolgi and at or near the cell surface, and short Aβ (Aβ40) is produced in the TGN/endosomal compartment and also at or near the cell surface.

Given an Aβ function as an axonal sorting signal of APP, release of Aβ may regulate the axonal transport of APP. Not only does the removal of the Aβ sequence from APP abolish axonal APP transport, but also free Aβ could – by blocking the APP binding site of the axonal transport machinery of APP – serve such a regulatory, physiological function. Excess intracellular and extracellular Aβ may convert the latter physiological function of Aβ to a pathogenic one by inhibiting the axonal transport of those proteins that use the same transport system as APP.

Because the apoEε4 allele may be associated with higher cholesterol levels in neurons, and because higher risk of developing Alzheimer's disease and axonal transport of membrane proteins are cholesterol dependent, we studied the influence of cholesterol on neuronal Aβ generation. By lowering the cholesterol level in neuronal cultures with statins (HMG-CoA reductase inhibitors), the formation

Research and Perspectives in Alzheimer's Diseases
Beyreuther/Christen/Masters (Eds.)
Neurodegenerative Disorders
© Springer-Verlag Berlin Heidelberg 2001

of secretory and intracellular Aβ is drastically reduced. Since the amount of Aβ produced by neurons is cholesterol dependent, both the physiological and pathogenic regulation of APP transport by Aβ appears to be controled in neurons by cholesterol. This finding implies a link between brain cholesterol. APP transport, Aβ production and the risk of developing Alzheimer's disease. These intriguing relationships open new strategies to influence the progression of Alzheimer's disease by modulating cholesterol biosynthesis of neurons with statins.

Introduction

A causal role for the 4-kDa β-amyloid (βA4/Aβ) peptide in the pathogenesis of Alzheimer's disease (AD) is supported by several recent findings. Mutations on chromosome 21 in the gene encoding the amyloid precursor protein (APP), have been linked to early-onset familial Alzheimer's disease (FAD). These mutations confer quantitative or qualitative changes in the production of APP-derived Aβ peptides, either the total amount of Aβ or the long form of Aβ ending in residue 42 (Aβ42; Ancolio et al. 1999; Cai et al. 1993; Citron et al. 1992; Eckman et al. 1997; Haass et al. 1994; Kwok et al. 2000; Suzuki et al. 1994). AD-causing APP missense mutations occur at or near the cleavage sites for the protease α-secretase, β-secretase and γ-secretase, respectively. The latter two release the N-terminus and C-terminus of Aβ, respectively, whereas the former excludes Aβ production by cleaving within the Aβ sequence. APP transgenic mice harboring one of these familial APP mutants exhibit amyloid pathology in the brain (Games et al. 1995; Hsiao et al. 1996; Johnson-Wood et al. 1997; Masliah et al. 1996; Sturchler-Pierrat et al. 1997). Down syndrome patients have an extra copy of the APP gene and develop AD 50 years earlier than people with normal karyotypes (Rumble et al. 1989). Importantly, two other identified genes on chromosome 14 and 1-presenilin 1 and presenilin 2, respectively are implicated in early-onset AD and also have been shown to increase Aβ production. In particular, the C-terminally elongated form of Aβ42 becomes elevated in vitro and in vivo by these AD causing missense mutations (Borchelt et al. 1996; Citron et al. 1997; Duff et al. 1996; Scheuner et al. 1996). Increased production and accumulation of Aβ42 appear to be the central events in the pathogenesis of the autosomal-dominant forms of FAD. Since the pathological lesions in the non-dominant inherited forms and in non-familiar forms of AD are very similar to those found in the autosomal dominant forms, all forms of AD may share common pathways in which the abnormal accumulation of Aβ42 plays a critical role. Indeed, Aβ pathology precedes the onset of disease by decades, indicating that it is a cause of the disease rather than a consequence (Rumble et al. 1989). However, amyloid plaques representing insoluble fibrillar Aβ-aggregates correlate poorly with the degree of damage in the end stage of AD (McLean et al. 1999). In contrast, in specific brain regions the loss of synaptophysin-immunoreactive presynaptic terminals (Terry et al. 1991) and the number of neurofibrillary tangles (Gomez-Isla et al. 1997) correlate well with cognitive decline in AD. A good correlation was found with the soluble Aβ levels or with total Aβ brain levels and the degree of cognitive impairment (Lue et al. 1999; McLean et al. 1999; Naslund et al. 2000; Wang et al. 1999). Furthermore, accumulation of Aβ in the form of SDS-stable Aβ dimers precedes neurofibrillary tangle formation in the CA1 region of the hippocampus (Funato et al. 1999). These findings support the amyloid cascade hypothesis of AD, which postulates that the accumulation of the Aβ peptide is the primary event in AD, preceding neurofibrillary tangle formation or neuronal loss in the cortex.

Currently much effort is focused on understanding the molecular basis of the Aβ-metabolism of APP, and therapeutic interventions in this pathway are considered promising in the design of rational therapy for, or prevention of, AD. While the proteases involved in the release of Aβ are intensely studied, β-secretase has been identified (Vassar et al. 1999) and an essential role of presenilin 1 for γ-secretase cleavage has been demonstrated (De Strooper et al. 1998; Li et al. 2000). Their physiological functions are largely unclear. Since the processing of APP, the signal transducing molecules Notch and Irel, and the transcription factor SREBP occurs in an analogous way (Brown et al. 2000; Niwa et al. 1999), it is logical to assume that the proteases involved regulate signal transduction mechanisms. Given that secretases regulate the physiological function of APP by producing free Aβ peptide, and given that AD mainly affects neurons, understanding the signal transduction mechanisms of APP and of its Aβ domain ultimately requires the study of APP expression, maturation, transport and metabolism in neurons. Here we briefly summarize our efforts to meet this challenge through the use of primary hippocampal and mixed cortical neurons of the rat by expressing human wild-type APP and variants thereof. The differences between the Aβ sequence of rodent and human APP make it possible to selectively evaluate the consequences of molecular manipulations on APP biogenesis, transport and metabolism (De Strooper et al. 1995; Simons et al. 1996; Tienari et al. 1996a, b, 1997). The availability of monoclonal antibodies to human Aβ (Ida et al. 1996a, b) also allows us to use electron microscopy to identify the specific cellular sites of Aβ production and to quantify Aβ production in vitro and vivo. In this way, we have been able to identify the physiological sites of Aβ production in neurons and to study the regulation of Aβ production by cholesterol. We were also able to analyze the consequences of deletions encompassing the Aβ region on APP trafficking and to determine the effects of amino acid exchanges within the membrane domain of APP on $A\beta_{42}$ production. In addition, we have expressed human APP in the developing wing of *Drosophila melanogaster* and thus obtained insight into the physiological function of APP. This review provides a general overview of these studies.

The APP Protein Family

The human APP cDNA was first cloned in 1987 (Kang et al. 1987) and other homologous genes were identified soon after (for references, see Coulson et al. 2000). The cDNA sequences for APP orthologues have been published for mon-

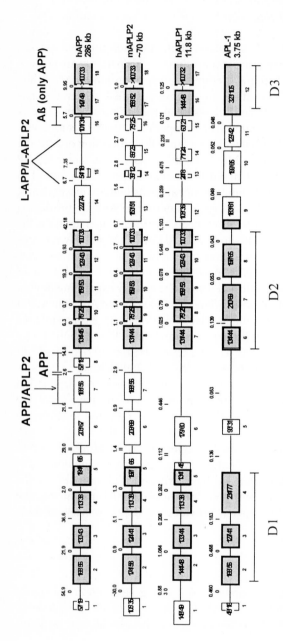

Fig. 1. Exon structure of known APP-family members (h, human; m, murine). Exons are indicated by rectangles and numbered from 5' to 3' (below each exon). The first number within an exon corresponds to nucleotides and the second to amino acid residues encoded by the exon. Reading frame changes by introns (0, I or II) and intron size (kilobases) are given above each intron. Exons or those parts of exons belonging to the three conserved domains (D1, D2 and D3, respectively) are framed in bold (Paliga K, PhD thesis, University of Heidelberg, 1997)

key, rat, mouse, guinea pig, chicken, frog (*Xenopus leavis*), two species of puffer fish (*Fugu rubripes* and *Tetradon fluviatilis*), zebra fish, and electric ray (*Narke japonica*). The amyloid sequences has also been determined for polar bear, pig, rabbit and sheep (Johnstone et al. 1991). APP-like cDNAs have been identified from worm (*C. elegans*) and fruitfly (*Drosophila melanogaster*), termed APL-1 and APPL, respectively (Daigle and Li 1993; Rosen et al. 1989). In addition, other members of the APP supergene family, the amyloid precursor-like proteins (APLPs), have also been isolated: APLP1 and APLP2 from human and mouse and the rat APLP2 sequence. This research showed that the human and murine APP supergene family comprises at least three members. The gene structure has been elucidated or deduced for human and *Fugu* APP, murine APLP1 and APLP2 and APL-1 of *C. elegans* (for references, see Coulson et al. 2000). Extensive search of the complete genomes of *C. elegans* and *D. melanogaster* did not result in any other gene products with significant homology to the known members of the APP supergene family, suggesting that there is only one family member in both *Caenorhabditis* and *Drosophila* (Coulson et al. 2000).

APL-1 and APPL are likely to be the ancestral genes to the other orthologues of the APP sugergene family members detected in mammals. So far, APP-like orthologues have not been described for unicellular organisms, and database searches of the yeast genomes of *Schizosaccharomyces pombe* and *Saccharomyces cerevisiae* did not reveal any putative orthologues of APP-like genes (Coulson et al. 2000).

A reconstruction of the possible evolution of the APP gene family suggested that the non-vertebrate genes APL-1 and APPL could not be the corresponding orthologues and functional homologues of APP. Between the invertebrate and mammalian APP homologues, there are obviously three conserved domains comprising exons 2–5 (cytoplasmic domain D1), 9–13 (cytoplasmic domain D2) and 17–18 (transmembrane domain D3) of APP (Coulson et al. 2000; Daigle and Li 1993; Lemaire et al. 1989; Fig. 1). These domains are retained within the super-gene family. The differences between the homologues include the gain of the KPI-domain corresponding to exon 7 in both APP and APLP2, the addition of the OX-2 homology exon 8 in human APP, and the sequence variability between the three domains D1-D3 (Fig. 1). The KPI-domain encodes a Kunitz type protease inhibitor and OX-2 stands for the MRC OX-2 gene, a lymphocyte proliferation marker.

In order to understand the evolution of the APP superfamily and to unravel the functional differences, phylogenetic trees were constructed that indicated a highly complex tree with at least three lineages (Fig. 2). First, the ancestral line-age with the genes from *C. elegans* and *D. melanogaster* appears to form a sepa-rate functional lineage. The second lineage includes APP found so far in amphibi-ans, birds, electric ray, primates, puffer fish and rodents, and the third group includes APLP1 as well as APLP2 isolated from rodents and primates. At least two gene duplication events are required to account for the evolution of the APP gene family: a APLP1 duplication subsequent to the speciation of insects and a later APLP1 duplication leading to vertebrate APP and APLP2 genes (Fig. 2).

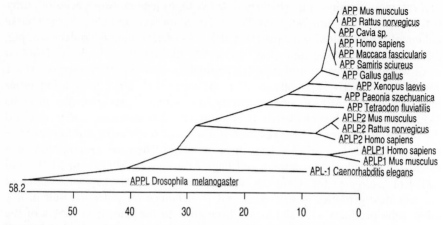

Fig. 2. Phylogenetic tree of APP and APP-like proteins. Algorithm for multiple alignment: clustal; residue weight table: PAM 250; multiple alignment parameters: gap penalty 10, gap length penalty 10 (Paliga K, PhD thesis, University of Heidelberg, 1997)

Proteins of the APP Supergene Family Have Adhesion Properties

All APP lineages retain some regions of homology that span the entire superfamily (Fig. 3). Among the highly conserved domains that are retained throughout the family is the zinc-binding domain (Bush et al. 1994; Multhaup et al. 1994), the collagen-binding domain (Beher et al. 1996) and the NPXY clathrin and adaptor protein attachment site (Koo and Squazzo 1994; Hynes 1999).

Conserved between APP and APLP2 is the chondroitin sulfate attachment site ENEGSG formed by alternative spliced exon 15 of human APP and exon 14 of APLP2 (Thinakaran and Sisodia 1994; Pangalos et al. 1995, Sandbrink et al. 1994a, b; Fig. 3). The heparin binding domain 1 is highly conserved in APP and APLP2 but absent in APLP1, APPL and APL-1, whereas the heparin binding domain 2 is present in all APP, APLP1 and APLP2 members (Fig. 3). Copper binding and reduction of copper(II) to copper(I) are mediated by APP and 40 % less efficiently by APLP2. The other members of the superfamily do not bind copper (Multhaup et al. 1996). The Aβ domain is a novel feature of APP, and therefore release of Aβ and its aggregation to amyloid fibrils and amyloid plaques are restricted to the APP lineage. These findings also illustrate that the functional differences within the APP superfamily do relate to the physiological as well as to the pathogenic function of the APP branch.

Physiological Function of APP

Although APP may not function as a true orthologue of APL-1, APPL, APLP1 or APLP2, there is evidence that vertebrate APP can function in invertebrates. Human APP was shown to partially recover the behavioral deficits of APPL null

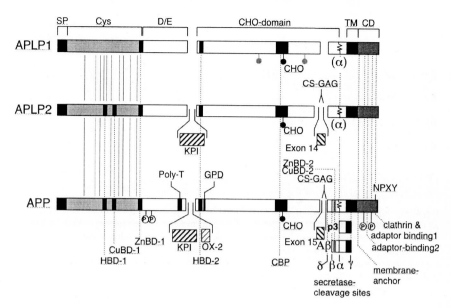

Fig. 3 Ligand binding map of the APP supergene family. SP, signal peptide; Cys, cysteine domain; D/E, acidic, asp- and glu-rich domain; poly T, threonine-rich site; CHO domain, N- and O-glyosylation domain; TM, transmembrane domain; CD, cytoplasmic domain; CS-GAG, ENEGSG-attachment site for xylosyl transferase resulting in chondroitin sulfate modification; KPI, Kunitz type protease inhibitor domain; OX-2, homology to lymphocyte proliferation marker MRC OX-2 gene; HBN-1/2, heparin-binding site one and two; CuBN-1/2, copper-binding site one and two; Zn-1/2, zinc-binding site one and two; GD, growth-promoting domain; CBP, collagen-binding domain secretase cleavage sites, APP processing sites; Aβ, β-amyloid domain of APP; P, phosphorylation site; NPXY, clathrin-binding and adaptor protein-binding site1; adaptor binding2, adaptor protein binding site 2; membrane-anchor, KKK sequence

D. melanogaster (Luo et al. 1992). Consistent with a role in cell adhesion is the finding that flies show a blistered wing phenotype due to a problem with cell contact between epithelial cells when human APP is expressed transgenically in the developing wing of *D. melanogaster* (Fossgreen et al. 1998; Fig. 4). A chimeric APP protein, with ectodomain, transmembrane and cytoplasmic sequences from APLP2 replacing the Aβ, transmembrane and cytoplasmic domains of APP, also showed a blistered wing phenotype, suggesting that all members of the family can mediate similar adhesive signals (Fig. 4).

As the wing of *D. melanogaster* is composed of interacting dorsal and ventral epithelial cell layers, this phenotype corroborates that human APP expression interferes with cell adhesion or adhesive signaling pathways independent of Aβ generation. Since transgenic expression of APP constructs that lack the cytoplasmic domain or both the cytoplasmic and transmembrane domains did not produce the blistered-wing phenotype, membrane association and the presence of a cytoplasmic domain of APP or APLP2 are necessary for the phenotype (Fossgreen et al. 1998; Fig. 4). The latter finding may indicate that a transmembrane signaling function of APP is imposed on the blistered-wing phenotype in a domi-

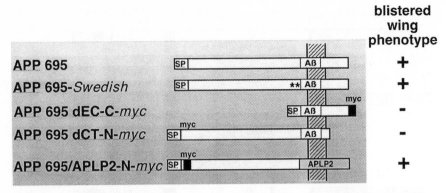

Fig. 4. The blistered-wing phenotype was produced by transgenic expression of APP695, APP695 carrying the *Swedish* mutations (**) (APP695-*Swedish*) and chimeric APP695 in which the Aβ, transmembrane and cytoplasmic domains of APP695 are replaced by the corresponding sequences from APLP2 (APP695/APLP2-N-*myc*). The blistered wings were not observed with the APP695 constructs lacking the entire ectodomain preceding the Aβ domain (APP695dEC-C-*myc*) or the entire cytoplasmic domain (APP695 dCT-N-*myc*). SP, APP signal peptide; N- or C-*myc*, N- or C-terminal position of a *myc* tag used to identify the corresponding gene products in the fly

nant function, presumably by interacting and inhibiting an appropriate partner on the cell membrane.

The blistered-wing phenotype resulting from APP expression in *Drosophila* may reflect a physiological function of APP in the regulation of cell adhesion and signaling. Adhesion properties are indeed conserved throughout the APP super-gene family, as mentioned before (Fig. 3). The ligand-binding characteristics make it apparent that the physiological function of APP is somehow related to the regulation of cell-cell or cell-substrate interactions.

Wing morphogenesis in *Drosophila* is characterized by the apposition of two epithelial cell layers, the dorsal and ventral layer, and may thus be used as a model to analyze the role of APP in the interaction of presynaptic and postsynaptic membranes at the synapse. We presume that APP negatively interacts with factors involved in the adhesion of the two wing epithelia and the two synaptic membranes, respectively.

In *Drosophila*, some mutations deficient for the expression of specific integrin subunits develop distinct wing blisters whereas another one was shown to be involved in short-term memory (Grotewiel et al. 1998). This provides an interesting link between physiological APP functions, integrin-mediated adhesion processes and memory mechanisms. Since integrins are heterodimeric receptors that may be ligand-bound to extracellular matrix molecules and may have a signaling function (Hynes 1999), integrins are especially required for the function of the cell-matrix-cell junction, where one surface of a cell such as a synaptic density adheres to the other. This function of integrins supports our suggestion of APP involvement in cell adhesion. We envisage that APP might act as an antagonist for integrins, both in the developing wing and at plasticity of the synapse. APP could bind, either directly or via other molecules (Fig. 3), to the same extracellular matrix sites as integrins. Indeed, in rat primary neurons, APP and integrins were colocalized in structures that are reminiscent of synapses (Storey et al. 1996). However, APP might be involved in other pathways, resulting in a deterioration of the adhesion or signaling pathways supported by the integrins. For example, expression of a mutant form of G protein involved in cAMP signaling or the blistered gene encoding a protein related to human serum response factors results in a comparable wing phenotype in *Drosophila* (Wolfgang et al. 1996; Prout et al. 1997). Taken together, *Drosophila* seems to be a valuable animal-model system for studying the physiological and pathogenic functions of human APP, since it may be suited for unraveling APP transmembrane signaling mechanisms potentially related to ADo.

Using this fly model to test genetic interactions between APP-expressing lines producing the blistered phenotype and other mutants engaged in wing morphology offers the perspective to unravel those physiological and pathogenic pathways linked to human APP (Fossgreen et al. 1998).

Physiological Function of the Aβ Domain of APP

Importantly and of etiological interest is the fact that the Aβ sequence is highly conserved between species from puffer fish to human, indicative of selective advantage and therefore function. Thus, despite its proposed causative role in AD, Aβ is likely to also be crucial for correct APP functioning, which may be specially important for regulation of axonal adhesion, neuritic outgrowth or regulation of synaptic plasticity. The latter gained support through the findings that in neurons APP has been detected in both pre- and postsynaptic sites, in vesicular elements in the cell body, axons and dendrites, as well as on the surface of axons and dendrites (Schubert et al. 1991; Simons et al. 1995). In these cells, APP undergoes fast anterograde and retrograde axonal transport (Koo et al. 1990; Kaether 2000).

In hippocampal neurons, APP is initially targeted to the axon and is only later found in the dendrites. This dendritic appearance may be due to transcytotic movement of APP from the axon to the dendrites (Simons et al. 1995; Yamazaki et al. 1995). The hierarchical appearance of newly synthesized APP, first in the axon and later in the dendrites, raises questions as to how this process is regulated. Ligand-induced transcytosis of the polymeric immunoglobulin receptor has been documented in epithelial cells and in neurons (Ikonen et al. 1993). Ligand binding could thus be one factor regulating the polarized distribution of APP in neurons. Which of the APP ligands summarized in Figure 3 are involved in the regulation of transcytosis remains to be elucidated. We suggest that copper is such a candidate since copper interferes with the homophilic interaction of APP and stabilizes the transmembrane form of APP (Beher et al. 1996; Borchardt et al. 1999).

When considering APP transcytosis as a sorting problem, it appears that neurons should use more than one sorting station in regulating polarized distri-

bution of APP. The first sorting station would be at the trans-Golgi network
(TGN), which is responsible for initial axonal delivery, and further sorting would
occur subsequently during endocytosis and transcytosis. When we analyzed the
sorting of APP using deletion mutants (Fig. 5), we observed three principal patterns of polarity: axonal, mixed and somatodendritic (Tienari et al. 1996a, b).
Thus the polarity could be reversed with some of the mutations, indicating that
APP contains both axonal and somatodendritic sorting signals that can be
sequentially decoded (Fig. 5).

When we tested the contribution of the cytoplasmic tail and the ectodomain
to the observed axonal delivery of APP, we found that a cytoplasmic tail deletion
mutant (dCT; Fig. 5) was distributed first to the axon. An ectodomain deletion
mutant SPC 99 (Fig. 5), extending from the beginning of the mature protein to
the beginning of the Aβ domain, showed a striking effect on the polarity of the
transport. For SPC 99, the protein polarity was virtually reversed with a predominantly somatodendritic localization. Expression of the ectodomain only as a soluble mutant (APPsec) also resulted in a somatodendritic distribution of the protein (Fig. 5). These results demonstrate that the ectodomain, but not the cytoplasmic tail, is required for axonal sorting of APP, and that APP ectodomain has
to be membrane-inserted for correct sorting. To map the ectodomain regions
needed for axonal sorting, we designed a set of deletions (dCys, dCHO and dAβ
in Fig. 5), that encompass 99 % of the amino acids in the APP ectodomain (Tienari et al. 1996a). The emergence of the somatodendritic pattern in the ectodomain deletion mutant SPC99 suggests that the recessive somatodendritic signals
are most likely located in the cytoplasmic tail of APP. The most significant effect
was detected with the deletion of the extracellular portion of Aβ (dAβ; Fig. 5),
which showed a somatodendritic predominance in its distribution. In contrast,
mutant SPC111, which includes the proximal part of exon 16 missing in SPC99
(Fig. 1), is again, at least partially, sorted to the axon. This finding suggests that
the conformation of the Aβ domain in SPC111 is similar to that of APP but different from that of SPC99. Surface labeling with antibodies showed that the trun-

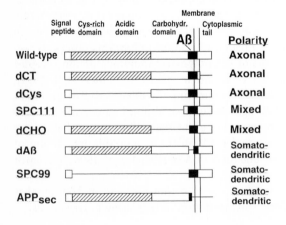

Fig. 5. Summary of APP deletion
mutants and their polarities in
hippocampal neurons. SP, APP signal peptide; dCys, deletion of the
cysteine-rich domain; dCHO, deletion of the N- and O-glycosylated
domain; dCT, deletion of the cytoplasmic domain; APPsec, APP
ectodomain corresponding to the
cleavage product by α-secretase;
dAβ, deletion of the ectodomain,
part of the Aβ domain; SPC99 and
SPC111, coexpression of SP with
C-terminal 99 or 111 residues of
APP

cated protein produced of SPC99 reached the plasma membrane, suggesting that the transmembrane topology of APP is maintained in the mutant dAβ. The profound effect on APP sorting found with deletion mutant dAβ suggests that the extracellular part of the Aβ domain would play an important role in axonal delivery and reveals a somewhat surprising physiological function for the Aβ domain of APP. An attractive hypothesis for the role of the Aβ domain in intraneuronal sorting would be that through its Aβ domain APP interacts directly with a sorting receptor in the trans-Golgi network (TGN) to get packaged into axonal transport vesicles.

Sites of Production for Aβ40 And Aβ42 Amyloid Peptides At Sorting Stations of APP

To identify directly the cellular compartments of Aβ generation, we have used two monoclonal antibodies that react with the Aβ neoepitopes produced by γ-secretase (Ida et al. 1996a). These antibodies recognize the free carboxyl-terminus of Aβ40 (X-40) and Aβ42 (X-42) and related peptides ending with Aβ residues 40 or 42, respectively, but not with APP or other APP fragments with different C-terminals (Ida et al. 1996a). Using these reagents and electron microscopy, one γ-secretase activity leading to Aβ40 was shown to reside in COS-7 and rat hippocampal neurons at the plasma membrane or at compartments very close to it (Hartmann et al. 1997). These observations are in good agreement with reports demonstrating immediate secretion of Aβ peptides into the conditioned medium after they are generated. Whereas Aβ40 was evenly distributed on the cell surface in COS-7 cells. Aβ42 peptides were mostly detected at cell-cell contact points and sites containing cellular processes (Hartmann et al. 1997). Intracellular labeling was very weak in COS-7 cells. The overall levels of Aβ40 immunoreactivity at the cell surface by far exceeded those of Aβ42. This finding is consistent with data obtained on Aβ40 and Aβ42 secretion into conditioned media (Suzuki et al. 1994). Our interpretation that γ-secretase is located at or near to the plasma membrane in these cells is supported by control experiments excluding the possibility that the plasma membrane labeling with our monoclonal antibodies specific for Aβ40 or Aβ42 is due to artifactually bound Aβ. In a previous study, we demonstrated lack of binding and uptake of Aβ by COS-7 cells from conditioned media (Ida et al. 1996b) as well as lack of binding and uptake of Aβ by primary hippocampal neurons from conditioned media (Hartmann et al. 1997).

Whereas nonneuronal cells have been shown to release Aβ and p3 peptides (Fig. 3) immediately upon its production, there is evidence that neuronal cells contain, intracellularly mainly the β-/γ-secretase product Aβ42 but not the α-/γ-secretase product p3 (Fuller et al. 1995; Turner et al. 1996; Tienari et al. 1997). We demonstrated that the intracellular and secreted Aβ species derive in neurons by distinct pathways in regard to ammonium chloride sensitivity and endocytosis dependence (Tienari et al. 1997). The secreted pool of Aβ is generated via an ammonium chloride-sensitive and endocytosis-dependent pathway, as has been shown in nonneuronal cells, whereas an alternative pathway generated the intracellular pool. The analysis of clinical APP mutants (APP$_{London}$ and AND APP$_{swe}$) in rat hippocampal neurons revealed that a common feature is their increased generation of Aβ42 (Tienari et al. 1997). The finding that two different APP mutations exhibit this same effect strengthens the hypothesis that Aβ42 would play a central role in AD. Furthermore, our results with APP$_{wild-type}$ and APP$_{London}$ indicate that Aβ42 might constitute a higher fraction intracellularly. Since neurons are the principal sites of damage in AD produce high levels of Aβ, and represent the only cell type with abundant intracellular Aβ, these results have important implications for AD. One of these implications may reside in the physiological function of the Aβ domain as an axonal targeting signal and of free Aβ as a putative inhibitor for the axonal transport of APP. As mentioned already, the latter would lead to an impaired axonal transport at least for those molecules that use the same transport receptors and transport molecules as APP.

Because Aβ42 seems to be the β-amyloid peptide affected by familial mutations in the APP and presenilin genes and thus the more disease-relevant secretory- as well as intracellular-accumulating isoform of neurons (Tienari et al. 1997; McLean et al. 1999), we analyzed the cellular compartments that produce Aβ42. We found that Aβ42 is produced very early in the sorting pathway in the endoplasmic reticulum (ER)/cis Golgi and the nuclear envelope that is continuous with the ER (Hartmann et al. 1997). Most intracellular labeling for Aβ40 was found in the TGN or vesicles in its vicinity and in late endosomes. Since no Aβ42 was detected in later intracellular compartments, including the TGN and late endosomes, and no Aβ40 was detected in the ER/cis Golgi and the nuclear envelope, the generation of the two Aβ species (Aβ40 and Aβ42), appears to occur in different subcellular compartments.

The specific intracellular production of Aβ in neurons presents a potential vulnerability to these cells, especially since intracellular Aβ production could lead to very high local Aβ concentrations (Tienari et al. 1997). High local concentrations of intracellular Aβ might increase its potential interference with the axonal transport machinery of APP. In addition, impaired transport of APP out of the ER/cisGolgi might enhance Aβ42 production and result in a pathogenic feedback loop between Aβ accumulation and APP retention at the sites of Aβ42 generation. Thus, Aβ might be deleterious to neurons even before it becomes secreted and its aggregation and deposition as senile plaques can be detected. This sequence would help to explain why neurons and especially pyramidal cells are highly vulnerable in AD and why Aβ deposition occurs only in the brain. Whether intracellular Aβ42 released from degenerating neurons contributes to senile plaque formation remains unclear.

Regulation of Neuronal Aβ Production By Cholesterol: Implication For Therapy

Cholesterol metabolism and AD are genetically linked. The apoEε4 allele of the apolipoprotein E gene is associated with higher cholesterol levels and has shown to increase the risk of developing the disease (Sing and Davignon 1985; Corder et al. 1993). The genetic link between cholesterol metabolism and AD led us to address the question of whether APP processing is cholesterol-dependent.

To study the influence of cholesterol on APP metabolism, we used primary cultures of rat hippocampal neurons infected with recombinant Semliki Forest virus carrying APP and APP mutants (Simons et al. 1998). This system has been used successfully to study the intracellular transport and processing of human APP (Simons et al. 1995, 1996; Tienari et al. 1996a, b, 1997). Inhibition of cholesterol biosynthesis and extraction of cholesterol from neuronal membranes were achieved by a combination of statin treatment and methyl-β-cyclodextrin extraction. Statins such as lovastatin or simvastatin allow in the presence of low amounts of mevalonate that control of the synthesis of cholesterol by inhibiting 3-hydroxy-3-methylglutaryl-CoA reductase (Alberts et al. 1980). Mevalonate is required for the synthesis of non-steroidal products and essential to preventing inhibition of the proteasome by lovastatin (Rao et al. 1999). β-Cyclodextrins have been shown to very efficiently and selectively extract cholesterol from the plasma membrane (Neufeld et al. 1996).

The combined treatment of rat hippocampal neurons with statins and methyl-β-cyclodextrin resulted in reduced Aβ secretion; however, APP levels and cell viability remained unaffected in both treated and control cells during the time course of the experiments (Simons et al. 1998). To demonstrate that the inhibition of secretion of Aβ by neurons was due to cholesterol synthesis inhibition and extraction from the plasma membrane and not to other effects of statins or methyl-β-cyclodextrin, two control experiments were performed. First, when cholesterol-loaded methyl-β-cyclodextrin was used, there was no reduction of Aβ secretion. Second, repletion of the cholesterol-depleted cells with cholesterol by adding back the cholesterol by methyl-β-cyclodextrin saturated with cholesterol fully restored the secretion of Aβ by neurons (Simons et al. 1998; Fig. 6). These data convincingly demonstrate that lowering of cholesterol by a combination of lovastatin and methyl-β-cyclodextrin reduces Aβ secretion in hippocampal neurons. On the other hand, cholesterol depletion of COS cells stably transfected with human APP695 did not lead to a significant decrease of Aβ production.

Since Aβ generation occurs in two steps (Fig. 3), we employed the two APP constructs APP695 and SPC99 (Fig. 5) to analyze which Aβ cleavage is inhibited by cholesterol-lowering drugs. The first cleavage of APP by β-secretase generates the 10-kDa fragment C99, which is further cleaved within the transmembrane domain by γ-secretase to produce Aβ (Fig. 3 and 5). The generation of APPsec by α-secretase (Fig. 3) leaves an 8-kDa transmembrane fragment in the cell membrane that is also further cleaved by γ-secretase to produce p3 (Fig. 3). When

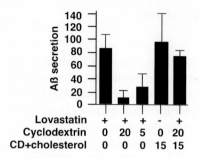

Fig. 6. Cholesterol lowering of hippocampal neurons reduces production and secretion of Aβ. Neurons were grown for four days in the presence (+) or absence (–) of lovastatin/mevalonate and after infection with SFV/APP were treated with 5 mM methyl-β-cyclodextrin for the indicated times (min). The relative secretion of Aβ is restored when cholesterol is added back for the indicated times (min) as a cholesterol-methyl-β-cyclodextrin (CD+cholesterol) inclusion

antibodies recognizing the C-terminal domain of APP were used to immunoprecipitate APP fragments from the cell lysates of cholesterol-depleted and control rat hippocampal neurons expressing human APP695, a dramatic inhibition of β-cleavage was detected, whereas the production of the fragment generated by α-sectretase was unperturbed. To analyze the effect of lovastatin treatment on the cleavage of γ-secretase, neurons expressing SPC99 were used, since removal of the transient signal sequence of SPC99 generates C99, which is identical with the C-terminal β-secretase product of APP. Conversion of C99 to Aβ requires γ-secretase. Both lovastatin treatment and methyl-β-cyclodextrin treatment – alone and together – of neurons expressing SPC99 inhibited Aβ production as well, suggesting that cholesterol-lowering regiments also inhibit γ-secretase cleavage (Bergmann and Hartmann, unpublished). Thus our combined results suggest that cholesterol is required for APP cleavage by β-secretase and γ-secretase but not for α-cleavage, which produces αAPPsec. We conclude that lowering of cholesterol affects amyloidogenic processing of APP while allowing nonamyloidogenic cleavage to proceed.

These findings raise the question of how cholesterol affects Aβ formation. One possibility is that reduction in membrane cholesterol changes the intracellular transport of APP so that the protein does not reach the cellular sites where β- and γ-secretase cleavage takes place. Alternatively, the β- and γ-secretase require cholesterol for their activity. We believe that the transport of APP is affected, since cholesterol depletion has been shown to affect transport of influenza virus hemagglutinin from the trans-Golgi network to the cell surface of non polarized baby hamster kidney cells as well as to the apical surface of polarized Madin-Darby canine kidney cells. The transport of the vesicular stomatitis virus (VSV) glycoprotein to the basolateral surface was not affected under these conditions (Ikonen and Simons 1998). These data strongly suggest that cholesterol serves as a sorting platform for inclusion of APP protein cargo destined for delivery to the apical membrane in nonneuronal cells and to the axonal membrane in neurons. Interestingly, conversion of the axonally transported glycosyl-phosphatidylinositol-anchored prion protein to the protease-resistant form PrPsc is reduced by cholesterol depletion (Taraboulos et al. 1995), suggesting that cholesterol is important for the generation of the disease-causing form of prions.

Conclusions

Many neurodegenrative diseases are associated with abnormal protein confor-mations leading to protein aggregation. Examples are AD, tauopathies, α-synucleinopathies such as Parkinson's disease, dementia with Lewy bodies and multiple system atrophy, Creutzfeldt-Jakob disease, amyotrophic lateral sclerosis, Huntington's disease and other diseases caused by polyglutamine expansion (Hardy and Gwinn-Hardy 1999; Goedert 1999; Price et al. 1998). Is AD a trans-port disease and as such a model to other chronic degenerative diseases of aging nervous system with the toxic gain of function of small, relatively insoluble, pro-tein aggregates that inhibit various cellular transport systems?

The finding that the Aβ domain of APP is essential for the axonal transport of APP implies that the Aβ domain or parts of it interact with a sorting receptor or a sorting platform for delivery of APP to the axonal membrane. This raises the question of how free Aβ affects brain function. One possibility is that excess free Aβ in neurons or in the neuropil might result in impaired APP transport due to an interaction with the sorting receptor or the sorting platform. Alternatively, given the tendency of Aβ to form higher aggregates, free Aβ could bind to the Aβ domain of APP. Both processes could interfere with the binding of APP to its sorting machinery and result in accumulation of APP at all sorting stations. This provides β- and γ-secretase with additional substrate. Thus, in a chronic cycle more and more Aβ40 and Aβ42 could be produced. Accumulation of intracellular and secreted Aβ might then interfere with axonal sorting and transcytosis of other proteins that use the same sorting machinery and platform as APP. Indeed, the accumulation of the axonally transported proteins tau – as paired helical fila-ments and neurofibrillary tangles – and α-synuclein – as fibrils within Lewy bod-ies – is frequently observed in AD. Since both pathologic lesions have been reported to coexist in brains of cases with familial APP mutations (Hardy 1994), it appears that the Aβ overproduction or accumulation in AD that interferes with axonal transport of tau and α-synuclein, respectively.

Since an intact axonal transport is vital to neurons, its impairment by Aβ may provide the missing link between APP, Aβ accumulation and neurodegener-ation. Regulation of APP transport as a means to control Aβ production becomes, therefore, a prime target for therapeutic research. How can APP trans-port be modulated? The axonal transport of transmembrane proteins or glycosyl-phosphatidylinositol-anchored proteins is cholesterol-dependent (Ikonen and Simons 1998). Since the Aβ domain of APP is essential for axonal transport of APP and cholesterol regulates Aβ production (Simons et al. 1998), this provides a link between APP transport, cholesterol biosynthesis and Aβ production. This intriguing relationship between cholesterol and Aβ production raises the hopes of new strategies to influence the progression of AD. The results summarized here suggest that cholesterol-lowering strategies might be beneficial to postpone the onset and perhaps also the progression of AD.

The brain has the highest cholesterol concentration in the body. However, very little cholesterol is taken up from circulating lipoproteins (Dietschy et al.

1993), suggesting that the majority of brain cholesterol is synthesized locally. Therefore, the usefulness of statins as a therapeutic approach for AD might largely depend on an efficient transfer of the corresponding statin through the blood-brain barrier. At present it remains unclear whether a low dosage of statin that can be tolerated in humans over a long period of time decreases cerebral Aβ42 and Aβ40 levels. In addition, it remains to be shown whether Aβ reduction alone is sufficient to prevent, postpone or ameliorate the disease.

Looking ahead, it is likely that a comprehensive package of genotypic analysis, presymptomatic diagnosis, and advice on preventive measures will be advocated with the use of drugs that effectively modify the course of the disease. Since many lines of evidence confirm that the generation of Aβ from APP ist the central pathway in AD, the following therapeutic strategies targeting the amyloidogenic APP-Aβ pathway appear most promising: inhibition of Aβ-forming secretases (Vassar et al. 1999; Li et al. 2000), redirection of APP processing away from Aβ (Borchardt et al. 1999; Simons et al. 1998), inhibition of Aβ aggregation (Hilbich et al. 1992), promotion of Aβ clearance (Bard et al. 2000; Vekrellis et al. 2000; Schenk et al. 2000), amelioration of Aβ toxicity and suppression of the reactive responses to Aβ toxicity (Christen 2000; Emilien et al. 2000; Fassbender et al. 2000). Future treatment will probably be based on combination therapies – such as neurotransmitter replacement combined with drugs directed against the targets mentioned above – tailored to the genetic profile of an individual (Masters and Beyreuther 1998). As exemplified for the regulation of Aβ production by cholesterol, the underlying concepts and principles for preventing the amyloidogenic processes from damaging neurons is straightforward and eminently amenable to intervention.

References

Alberts AW, Chen J, Kuron G, Hunt V, Huff J, Hoffman C, Rothrock J, Lopez M, Joshua H, Harris E, Patchett A, Monaghan R, Currie S, Stapley E, Albers-Schonberg G, Hensens O, Hirshfield J, Hoogsteen K, Liesch J, Springer J (1980) Mevinolin: a highly potent competitive inhibitor of hydroxymethylglutaryl-coenzyme A reductase and a cholesterol-lowering agent. Proc Natl Acad Sci USA 77: 3957–3961

Ancolio K, Dumanchin C, Barelli H, Warter JM, Brice A, Campion D, Frébourg T, Checkler F (1999) Unusual phenotypic alteration of beta amyloid precursor protein (beta APP) maturation by a new Val715→Met betaAPP-770 mutations responsible for probable early-onset Alzheimer disease. Proc Natl Acad Sci USA 96: 4119–4124

Bard F, Cannon C, Barbour R, Burke RL, Games D, Grajeda H, Guido T, Hu K, Huang J, Johnson-Wood K, Khan K, Kholodenko D, Lee M, Lieberburg I, Motter R, Nguyen M, Soriano F, Vasquez N, Weiss K, Welch B, Seubert P, Schenk D, Yednock T (2000) Peripherally administered antibodies against amyloid beta-peptide enter the central nervous system and reduce pathology in a mouse model of Alzheimer disease. Nat Med 6: 916–919

Beher D, Hesse L, Master CL, Multhaup G (1996) Regulation of amyloid protein precursor (APP) binding to collagen and mapping of the binding sites on APP and collagen type I. J Biol Chem 271: 1613–1620

Borchardt T, Camakaris J, Cappai R, Masters CL, Beyreuther K, Multhaup G (1999) Copper inhibits beta-amyloid production and stimulates the non-amyloidogenic pathway of amyloid-precursor-protein secretion. Biochem J 344: 461–467.

Borchelt DR, Thinakaran G, Eckman CB, Lee MK, Davenport F, Ratovisky T, Prada CM, Kim G, See-kins S, Yager D, Slunt HH, Wang R, Seeger M, Levey M, Levey AI, Gandy SE, Copeland NG, Jen-kins NA, Price DL, Younkin SG, Sisodia SS (1996) Familial Alzheimer's disease-linked presenilin 1 variants elevate Aβ1-42/1-40 ratio in vitro and in vivo. Neuron 17: 1005–1013

Brown MS, Ye J, Rawson RB, Goldstein JL (2000) Regulated intramembrane proteolysis: a control mechanism conserved from bacteria to humans. Cell 100: 391–398

Bush AI, Pettingell WJ, de Paradis M, Tanzi R, Wasco W (1994) The amyloid beta-protein precursor and its mammalian homologues. Evidence for a zinc-modulated heparin-binding superfamily. J Biol Chem 269: 618–621

Cai XD, Golde TE, Younkin SG (1993) Release of excess amyloid beta protein from a mutant amyloid beta protein precursor. Science 259: 514–516

Christen Y (2000) Oxidative stress and Alzheimer disease. Am J Clin Nutr 71: 621S–629S

Citron M, Oltersdorf T, Haass C, McConlogue L, Hung AY, Seubert P, Vigo-Pelfrey C, Lieberburg I, Sel-koe D (1992) Mutation of the beta-amyloid precursor protein in familial Alzheimer's disease increases beta-protein production. Nature 360: 672–674

Citron M, Westaway D, Xia W, Carlson G, Diehl T, Levesque G, Johnson-Wood K, Lee M, Seubert P, Davis A, Kholodenko D, Motter R, Sherrington R, Perry B, Yao H, Strome R, Lieberburg I, Rom-mens J, Kim S, Schenk D, Fraser P, St George Hyslop P, Selkoe DJ (1997) Mutant presenilins of Alz-heimer's disease increase production of 42-residue amyloid β-protein in both transfected cells and transgenic mice. Nat Med 3: 67–68

Corder EH, Saunders AM, Strittmatter WJ, Schmechel DE, Gaskell PC, Small GW, Roses AD, Haines JL, Pericak-Vance MA (1993) Gene dose of apolipoprotein E type 4 allele and the risk of Alzheimer's dis-ease in late onset families. Science 261: 921–923

Coulson EJ, Paliga K, Beyreuther K, Masters CL (2000) What the evolution of the amyloid precursor supergene family tells us about its function. Neurochem Int 36: 175–184

Daigle I, Li C (1993) apl-1, A Caenorhabditis elegans gene encoding a protein related to the human beta-amyloid protein precursor. Proc Natl Acad Sci USA 90: 12045–12049

De Strooper B, Simons M, Multhaup G, Van Leuven F, Beyreuther K, Dotti CG (1995) Production of intracellular amyloid-containing fragments in hippocampal neurons expressing human amyloid pre-cursor protein and protection against amyloidogenesis by subtle amino acid substitutions in the rodent sequence. EMBO J 14: 4932–4938

De Strooper B, Saftig P, Craessaerts K, Vanderstichele H, Guhde G, Annaert W, Von Figura K, Van Leu-ven F (1998) Deficiency of presenilin-1 inhibits the normal cleavage of amyloid precursor protein. Nature 391: 387–390

Dietschy JM, Turley SD, Spady DK (1993) Role of liver in the maintenance of cholesterol and low density lipoprotein homeostasis in different animal species, including humans. J Lipid Res 34: 1637–1659

Duff K, Eckman C, Zehr C, Yu X, Prada CM, Perez Tur J, Hutton M, Buee L, Harigaya Y, Yager D, Mor-gan D, Gordon MN, Holcomb L, Refolo L, Zenk B, Hardy J, Younkin S (1996) Increased amyloid-β42(43) in brains of mice expressing mutant presenilin 1. Nature 383: 710–713

Eckman CB, Mehta ND, Crook R, Perez-tur J, Prihar G, Pfeifer E, Graff-Radford N, Hinder P, Yager D, Zenk B, Refolo LM, Prada CM, Younkin SG, Hutton M, Hardy J (1997) A new pathogenic mutation in the APP gene (I716V) increases the relative proportion of A beta 42(43). Human Mol Genet 6: 2087–2089

Emilien G, Beyreuther K, Masters CL, Maloteaux JM (2000) Prospects for pharmacological interven-tion in Alzheimer disease. Arch Neurol 57: 454–459

Fassbender K, Masters C, Beyreuther K (2000) Alzheimer's disease: An inflammatory disease? Neuro-biol Aging 21: 433–436

Fossgreen A, Bruckner B, Czech C, Masters CL, Beyreuther K, Paro R (1998) Transgenic Drosophila expressing human amyloid precursor protein show gamma-secretase activity and a blistered wing phenotype. Proc Natl Acad Sci USA 95: 13703–13708

Fuller SJ, Storey E, Li QX, Smith AI, Beyreuther K, Masters CL (1995) Intracellular production of beta A4 amyloid of Alzheimer's disease: modulation by phosphoramidon and lack of coupling to the secretion of the amyloid precursor protein. Biochemistry 34: 8091–8098

Funato H, Enya M, Yoshimura M, Morishima-Kawashima M, Ihara Y (1999) Presence of sodium dode-cyl sulfate-stable amyloid beta-protein dimers in the hippocampus CA1 not exhibiting neurofibril-lary tangle formation. Am J Pathol 155: 23–28

Games D, Adams D, Alessandrini R, Barbour R, Berthelette P, Blackwell C, Carr T, Clemens J, Donald-son T, Gillespie F, Guido T, Hagopian S, Johnson-Wood K, Khan I, Lee M, Leibowitz P, Lieberburg I, Little S, Masliah E, McConlogue L, Montoya Azvala M, Mucke L, Paganini L, Penniman E, Power M, Schenk D, Seubert P, Snyder B, Soriano F, Tan H, Vitale J, Wadsworth S, Wolozin B, Zhao J (1995) Alzheimer-type neuropathology in transgenic mice overexpressing V717F β-amyloid precursor pro-tein. Nature 373: 523–527

Goedert M (1999) Filamentous nerve cell inclusions in neurogenerative diseases: tauopathies and alpha-synucleinopathies. Phil Trans R Soc Lond B Biol Sci 354: 1101–1118

Gomez-Isla T, Hollister R, West H, Mui S, Growdon JH, Petersen RC, Parisi JE, Hyman BT (1997) Neuronal loss correlates with but exceeds neurofibrillary tangles in Alzheimer's disease. Ann Neurol 41: 17–24

Grotewiel MS, Beck CD, Wu KH, Zhu XR, Davis LD (1998) Integrin-mediated short-term memory in Drosophila. Nature 391: 455–460

Haass C, Hung AY, Selkoe DJ, Teplow DB (1994) Mutations associated with a locus for familial Alzhei-mer's disease result in alternative processing of amyloid beta-protein precursor. J Biol Chem 269: 17741–17748

Hardy J (1994) Lewy bodies in Alzheimer's disease in which the primary lesion is a mutation in the amyloid precursor protein. Neurosci Lett 180: 290–291

Hardy J, Gwinn-Hardy K (2000) Neurodegenerative disease: a different view of diagnosis. Mol Med Today 5: 514–517

Hartmann T, Bieger SC, Brühl B, Tienari PJ, Ida N, Allsop D, Roberts GW, Masters CL, Dotti CG, Unsicker K, Beyreuther K (1997) Distinct sites of intracellular production for Alzheimer's disease Aβ40/42 amyloid peptides. Nature Med 3: 1016–1020

Hilbich C, Kisters Woike B, Reed J, Masters CL, Beyreuther K (1992) Substitutions of hydrophobic amino acids reduce the amyloidogenicity of Alzheimer's disease beta A4 peptides. J Mol Biol 228: 460–473

Hynes RO (1999) Cell adhesion: old and new questions. Trends Cell Biol 9: M33–M37

Hsiao K, Chapman P, Nilsen S, Eckman C, Harigaya Y, Younkin S, Yang FS, Cole G (1996) Correlative memory deficits, Aβ elevation and amyloid plaques in transgenic mice. Science 274: 99–102

Ida N, Hartmann T, Pantel J, Schroder J, Zerfass R, Forstl H, Sandbrink R, Masters CL, Beyreuther K (1996a) Analysis of heterogeneous A4 peptides in human cerebrospinal fluid and blood by a newly developed sensitive Western blot assay. J Biol Chem 271: 22908–22914

Ida N, Masters CL, Beyreuther K (1996b) Rapid cellular uptake of Alzheimer amyloid βA4 peptide by cultured human neuroblastoma cells. FEBS Lett 394: 174–178

Ikonen E, Simons K (1998) Protein and lipid sorting from the trans-Golgi network to the plasma mem-brane in polarized cells. Semin Cell Dev Biol 9: 503–509

Ikonen E, Parton RG, Hunziker W, Simons K, Dotti C (1993) Transcytosis of the polymeric immuno-globin receptor in hippocampal neurons. Curr Biol 3: 635–644

Johnson-Wood K, Lee M, Motter R, Hu K, Gordon G, Barbour R, Khan K, Gordon M, Tan H, Games D, Lieberburg I, Schenk D, Seubert P, McConlogue L (1997) Amyloid precursor protein processing and Aβ42 deposition in a transgenic mouse model of Alzheimer's disease. Proc Natl Acad Sci USA 94: 1550–1555

Johnstone EM, Chaney MO, Norris FH, Pascual R, Little SP (1991) Conservation of the sequence of the Alzheimer's disease amyloid peptide in dog, polar bear and five other mammals by cross-species polymerase chain reaction analysis. Brain Res Mol Brain Res 10: 299–305

Kaether C, Skehel P, Dotti CG (2000) Axonal membrane proteins are transported in distinct carriers: a two-color video microscopy study in cultured hippocampal neurons. Mol Biol Cell 11: 1213–1224

Kang J, Lemaire HG, Unterbeck A, Salbaum JM, Masters CL, Grzeschik KH, Multhaup G, Beyreuther K, Müller-Hill B (1987) The precursor of Alzheimer's disease amyloid A4 protein resembles a cell-surface receptor. Nature 325, 733–736

Koo EH, Squazzo SL (1994) Evidence that production and release of amyloid beta-protein involves the endocytic pathway. J Biol Chem 269: 17386–17389

Koo EH, Sisodia SS, Archer DR, Martin LJ, Weidemann A, Beyreuther K, Fischer P, Masters CL, Price DL (1990) Precursor of amyloid protein in Alzheimer disease undergoes fast anterograde axonal transport. Proc Natl Acad Sci USA 87: 1561–1565

Kwok JB, Li QX, Hallup M, Whyte S, Ames D, Beyreuther K, Masters CL, Schoffield PR (2000) Novel Leu723Pro amyloid precursor protein mutation increases amyloid beta42(43) peptide levels and induces apoptosis. Ann Neurol 47: 249–253

Lemaire HG, Salbaum JM, Multhaup G, Kang J, Bayney RM, Unterbeck A, Beyreuther K, Muller Hill B (1989) The PreA4(695) precursor protein of Alzheimer's disease A4 amyloid is encoded by 16 exons. Nucleic Acids Res 25: 517–522

Li YM, Xu M, Lai MT, Huang Q, Castro JL, DiMuzio-Mower J, Harrison T, Lellis C, Nadin A, Neduveilli TG, Reguster RB, Sardana MK, Shearman MS, Smith AL, Shi XP, Yin KC, Shafer JA, Gardell ST (2000) Photoactivated γ-secretase inhibitors directed to the active site covalently label presenilin 1. Nature 405: 689–694

Lue LF, Kuo YM, Roher AE, Brachova L, Shen Y, Sue L, Beach T, Kurth JH, Rydel RE, Rogers J (1999) Soluble amyloid beta peptide concentration as a predictor of synaptic change in Alzheimer's disease. Am J Pathol 155: 853–862

Luo L, Tully T, White K (1992) Human amyloid precursor protein ameliorates behavioral deficit of flies deleted for Appl gene. Neuron 9: 595–605

Masliah E, Sisk A, Mallory M, Mucke L, Schenk D, Games D (1996) Comparison of neurodegenerative pathology in transgenic mice overexpressing V717F β-amyloid precursor protein and Alzheimer's disease. J Neurosci 16: 5795–5811

Masters CL, Beyreuther K (1998) Alzheimer's disease. Brit Med J 316: 446–448

McLean CA, Cherry RA, Fraser FW, Fuller SJ, Smith MJ, Beyreuther K, Bush AI, Masters CL (1999) Soluble pool of Aβ as a determinant of severity of neurodegeneration in Alzheimer's disease. Ann Neurol 46: 860–866

Multhaup G, Bush AI, Pollwein P, Masters CL (1994) Interaction between the zine (II) and the heparin binding site of the Alzheimer's disease beta A4 amyloid precursor protein (APP). FEBS Lett 355: 151–154

Multhaup G, Schlicksupp A, Hesse L, Beher D, Ruppert T, Masters CL, Beyreuther K (1996) The amyloid precursor protein of Alzheimer's disease in the reduction of copper(II) to copper(I). Science 271: 1406–1409

Naslund J, Haroutunian V, Mohs R, Davis KL, Davies P, Greengard P, Buxbaum J (2000) Correlation between elevated levels of amyloid beta-peptide in the brain and cognitive decline JAMA 283: 1571–1577

Neufeld EB, Cooney AM, Pitha J, Dawidowicz EA, Dwyer NK, Pentchev PG, Blanchette-Mackie EJ (1996) Intracellular trafficking of cholesterol monitored with a cyclodextrin. J Biol Chem 271: 21604–21613

Niwa M, Sidrauski C, Kaufman RJ, Walter P (1999) A role for presenilin-1 in nuclear accumulation of Ire1 fragments and induction of the mammalian unfolded protein response. Cell 99: 691–702

Pangalos MN, Efthimiopoulos S, Shioi J, Robakis NK (1995) The chondroitin sulfate attachment site of appican is formed by splicing out exon 15 of the amyloid precursor gene. J Biol Chem 270: 10388–10391

Price DL, Sisodia SS, Borchelt DR (1998) Genetic neurodegenerative diseases: the human illness and transgenic models. Science 282: 1079–1083

Prout M, Damania Z, Soong J, Fristrom D, Fristrom JW (1997) Autosomal mutations affecting adhesion between wing surfaces in Drosophila melanogaster. Genetics 146: 275–285

Rao S, Porter DC, Chen X, Herliczek T, Lowe M, Keyomarsi K (1999) Lovastatin-mediated G1 arrest is through inhibition of the proteasome, independent of hydroxymethyl glutaryl-CoA reductase. Proc Natl Acad Sci USA 96: 7797–7802

Rosen DR, Martin-Morris L, Luo LQ, White K (1989) A Drosophila gene encoding a protein resembling the human beta-amyloid protein precursor. Proc Natl Acad Sci USA 86: 2478–2482

Rumble B, Retallack R, Hilbich C, Simms G, Multhaup G, Martins R, Hockey A, Montgomery P, Beyreuther K, Masters CL (1989) Amyloid A4 protein and its precursor in Down's Syndrome and Alzheimer's disease. N Engl J Med 320: 1446–1452

Sandbrink R, Masters CL, Beyreuther K (1994a) APP gene family: unique age-associated changes in splicing of Alzheimer's betaA4-amyloid protein precursor. Neurobiol Dis 1: 13–24

Sandbrink R, Masters CL, Beyreuther K (1994b) Similar alternative splicing of a non-homologous domain in beta A4-amyloid protein precursor-like proteins. J Biol Chem 269: 14227–14234

Schenk DB, Seubert P, Lieberburg I, Wallace J (2000) Beta-peptide immunization: a possible new treatment for Alzheimer disease. Arch Neurol 57: 934–936

Scheuner D, Eckman C, Jensen M, Song X, Citron M, Suzuki N, Bird TD, Hardy J, Hutton M, Kukull W, Larson E, Levey-Lahad E, Viitanen M, Peskind E, Poorkaj P, Schellenberg G, Tanzi R, Wasco W, Lannfelt L, Selkoe D, Younkin S (1996) Secreted amyloid β-protein similar to that in the senile plaques of Alzheimer's disease is increased in vivo by the presenilin 1 and 2 and APP mutations linked to familial Alzheimer's disease. Nat Med 2: 864–870

Schubert W, Prior R, Weidemann A, Dircksen H, Multhaup G, Masters CL, Beyreuther K (1991) Localization of Alzheimer beta A4 amyloid precursor protein at central and peripheral synaptic sites. Brain Res 563: 184–194

Simons M, Ikonen E, Tienari PJ, Cid-Arregui A, Monning U, Beyreuther K, Dotti CG (1995) Intracellular routing of human amyloid protein precursor: axonal delivery followed by transport to the dendrites. J Neurosci Res 41: 121–128

Simons M, de Strooper B, Multhaup G, Tienari PJ, Dotti CG, Beyreuther K (1996) Amyloidogenic processing of the human amyloid precursor protein in primary cultures of rat hippocampal neurons. J Neurosci 16: 899–908

Simons M, Keller P, De Strooper B, Beyreuther K, Dotti CG, Simons K (1998) Cholesterol depletion inhibits the generation of beta-amyloid in hippocampal neurons. Proc Natl Acad Sci USA 95: 6460–6464

Sing CF, Davignon J (1985) Role of the apolipoprotein E polymorphism in determining normal plasma lipid and lipoprotein variation. Am J Human Genet 37: 268–285

Storey E, Beyreuther K, Masters CL (1996) Alzheimer's disease amyloid precursor protein on the surface of cortical neurons in primary culture co-localizes with adhesion patch components. Brain Res 735: 217–231

Sturchler-Pierrat C, Abramowski D, Duke M, Wiederhold KH, Mistl C, Rothacher S, Ledermann B, Burki K, Frey P, Paganetti PA, Waridel C, Calhoun ME, Jucker M, Probst A, Staufenbiel M, Sommer B (1997) Two amyloid precursor protein transgenic mouse models with Alzheimer disease-like pathology. Proc Natl Acad Sci USA 94: 13287–92

Suzuki N, Cheung TT, Cai XD, Odaka A, Otvos L, Eckman Jr C, Golde TE, Younkin SG (1994) An increased percentage of long amyloid beta protein secreted by familial amyloid beta protein precursor (beta APP717) mutants. Science 264: 1336–1340

Taraboulos A, Scott M, Semenov A, Avrahami D, Laszlo L, Prusiner SB, Avraham D (1995) Cholesterol depletion and modification of COOH-terminal targeting sequence of the prion protein inhibit formation of the scrapie isoform. J Cell Biol 129: 121–132

Terry RD, Masliah E, Salmon DP, Butters N, DeTeresa R, Hill R, Hansen LA, Katzman R (1991) Physical basis of cognitive alterations in Alzheimer's disease: synapse loss is the major correlate of cognitive impairment. Ann Neurol 30: 572–580

Tienari PJ, De Strooper B, Ikonen E, Ida N, Simons M, Masters CL, Dotti CG, Beyreuther K (1996a) Neuronal sorting and processing of amyloid precursor protein: implications for Alzheimer's disease. Cold Spring Harb Symp Quant Biol 61: 575–585

Tienari PJ, De Strooper B, Ikonen E, Simons M, Weidemann A, Czech C, Hartmann T, Ida N, Multhaup G, Masters CL, Van Leuven F, Beyreuther K, Dotti CG (1996b) The beta-amyloid domain is essential for axonal sorting of amyloid precursor protein. EMBO J 15: 5218–5229

Tienari PJ, Ida N, Ikonen E, Simons M, Weidemann A, Multhaup G, Masters CL, Dotti CG, Beyreuther K (1997) Intracellular and secreted Alzheimer beta-amyloid species are generated by distinct mechanisms in cultured hippocampal neurons. Proc Natl Acad Sci USA 94: 4125–4130

Thinakaran G, Sisodia SS (1994) Amyloid precursor-like protein 2 (APLP2) is modified by the addition of chondroitin sulfate glycosaminoglycan at a single site. J Biol Chem 269: 22099–22104

Trommsdorff M, Borg JP, Margolis B, Herz J (1998) Interaction of cytosolic adaptor proteins with neuronal apolipoprotein E receptors and the amyloid precursor protein. J Biol Chem 273: 33556–33560

Turner RS, Suzuki N, Chyung AS, Younkin SG, Lee VM (1996) Amyloids beta40 and beta42 are generated intracellularly in cultured human neurons and their secretion increases with maturation. J Biol Chem 271: 8966–8970

Vassar R, Bennett BD, Babu-Khan S, Kahn S, Mendiaz EA, Denis P, Teplow DB, Ross S, Amarante P, Loeloff R, Luo Y, Fisher S, Fuller J, Edenson S, Lile J, Jarosinski MA, Biere AL, Curran E, Burgess T, Louis JC, Collins F, Treanor J, Rogers G, Citron M (1999) Beta-secretase cleavage of Alzheimer's amyloid precursor protein by the transmembrane aspartic protease BACE. Science 286: 735–741

Vekrellis K, Ye Z, Qui WQ, Walsh D, Hartley D, Chesneau V, Rosner MR, Selkoe DJ (2000) Neurons regulate extracellular levels of amyloid beta-protein via proteolysis by insulin-degrading enzyme. J Neurosci 20: 1657–1665

Wang J, Dickson DW, Trojanowski JQ, Lee VM (1999) The levels of soluble versus insoluble brain Abeta distinguish Alzheimer's disease from normal and pathologic aging. Exp Neurol 158: 328–337

Wolfgang WJ, Roberts IJ, Quan F, O'Kane C, Forte M (1996) Activation of protein kinase A-independent pathways by Gs alpha in Drosophila. Proc Natl. Acad Sci USA 93: 14542–14547

Yamazaki T, Selkoe DJ, Koo EH (1995) Trafficking of cell surface beta-amyloid precursor protein: retrograde and transcytotic transport in cultured neurons. J Cell Biol 129: 431–442

The NEXT Step in Notch Processing and its Relevance to Amyloid Precursor Protein

R. Kopan, S. Huppert, J. S. Mumm, M. T. Saxena, E. H. Schroeter,
W. J. Ray, and A. Goate

The last Nobel Prize in Physiology and Medicine awarded in the twentieth century went to Günter Blobel for his discovery that proteins have intrinsic signals governing their transport and localization in the cell. It is only fitting that, at the close of the millennium, the confluence of several unrelated fields resulted in the emergence of a new paradigm for signal transduction: regulated intramembranous proteolysis (RIP: Brown et al. 2000) of "dual address" proteins. Scientists working in topics as unrelated as Alzheimer's disease, bacterial sporulation, lipid metabolism, Notch signaling and unfolded protein response all contributed to this realization. "Dual address" proteins contain two intrinsic signals: the first directs proteins to a holding site where – in response to stimulus – they undergo intramembranous proteolysis, releasing a subdomain that rides a second intrinsic signal to its site of action, often the nucleus. To researchers in the Alzheimer's field, this realization provides new insight into the biological function of presenilin, which was discovered in humans due to its involvement in familial Alzheimer's disease (FAD) and in the worm *C. elegans* for its role in Notch signaling. This article will describe these developments from the perspective of the Notch protein, in particular the advances that have occurred over the last five years in the Notch field. This period was critical in shaping our current understanding of how a signal is transduced through the Notch receptor. Defining Notch as a substrate for presenilin, and the emergence of additional substrates, has helped elucidate the role of presinilin proteins in the cell and will assist in the search for a treatment for Alzheimer's disease.

Notch Structure and Function: Overview

Notch proteins, first described in *Drosophila* (for a historical perspective, see Wu and Rao 1999) are large ($\sim$ 2500 aminoacids), type one transmembrane receptors with an extracellular domain containing up to 36 EGF repeats and a membrane-proximal region containing lin-Notch repeats (LNR) and two conserved cysteines (Greenwald 1994). The extracellular domain undergoes a Furin-convertase type cleavage in the secretory pathway (Blaumueller et al. 1997; Logeat et al. 1998); the resulting polypeptide chains are held together by calcium coordinating bonds (Rand et al. 2000). When Notch is present at the plasma membrane, the Notch intracellular domain (NICD) is thought to be inert, but

Research and Perspectives in Alzheimer's Diseases
Beyreuther/Christen/Masters (Eds.)
Neurodegenerative Disorders
© Springer-Verlag Berlin Heidelberg 2001

some unusual features provide clues that Notch is a dual address protein. NICD contains nuclear localizing signals, a multitude of protein-protein interaction domains (the most famous of which are the Ankyrin repeats) and a cluster of charged amino acids (PEST and OPA repeats) at the C terminus that are often found in a transcriptional activation domain.

Notch receptors are activated by type on transmembrane ligands known collectively as DSL proteins (*Delta, Serrate* and *Lag 2* are the founding members of this family) and is therefore involved in short-range cell-cell interactions. Notch-mediated signals permit equivalent cells to acquire the proper fate during development in many metazoans (reviewed in Angerer and Angerer 1999; Bray 1998; Greenwald 1998; Joutel and Tournier-Lasserve 1998; Kimble et al. 1998; Lewis 1998b; Miele and Osborne 1999; Robey 1999; Rooke and Xu 1998; Saito and Watanabe 1998). Neurogenesis in *Drosophila* is a well-studied system where Notch-mediated cell-cell communications are critical in establishing the proper allocation of cells to the neuronal fate (Artavanis-Tsakonas and Simpson 1991; Artavanis-Tsakonas et al. 1999). Clusters of ectodermal cells acquire neurocompetence due to induction of *acheate-scute* complex genes by upstream signaling events. Notch-mediated lateral interactions resolve this cluster into a group of epidermal cells with one neuronal progenitor located in a stereotypical position within the cluster. Observations of mosaic clones in the *Drosophila* ectoderm that traverse a proneural cluster, such that cells with different genotypes are in direct contact, revealed that marked wild type cells stochastically select the neuronal fate. However, if neighboring cells contain different dosages of *Notch*, the cell with fewer copies will always differentiate as a neuronal progenitor and the cell with more copies of *Notch* will always differentiate as an epidermal cell. These experiments demonstrate that cells can sense the dosage of the *Notch* genes on their neighbors, possibly due to subtle differences in the surface density of the Notch protein (Heitzler and Simpson 1991, 1993). Our current understanding of the mechanism allowing cells to accomplish this feat is illustrated in recent reviews (Collier et al. 1996; Kopan and Turner 1996; Lewis 1996, 1998a, b; Posakony 1994). Equivalent cells begin with sub-threshold levels of transcription factors present in their nuclei; these transcription factors can divert cells into different fates. The cells also contain a nuclear DNA binding protein called CSL (an acronym for CBF in vertebrates, Suppressor of hairless in flies and Lag-1 in *C. elegans*). The CSL protein associates with transcriptional co-oppressors (Hsieh et al. 1999; Kao et al. 1998) and is thought to act as a transcriptional repressor in the absence of additional signal (Hsieh et al. 1996; Hsieh and Hayward 1995; Morel and Schweisguth 2000; Olave et al. 1998). These cells also express equivalent amounts of the Notch receptor and its ligand(s). As the cells interact, ligand binding to Notch triggers intramembranous proteolysis of Notch (Kidd et al. 1998; Lecourtois and Schweisguth 1998; Schroeter et al. 1998; Struhl and Adachi 1998). Either because one of the two interacting cells has a higher density of Notch, or because this cell receives less antagonizing input or more promoting impact from other signaling pathways, or just stochastically, one cell will accumulate NICD in its nucleus to reach a threshold level first. This cell will convert the CSL protein into a transcriptional activator of downstream genes. Among the targets of the CSL/Notch complex is Notch itself and transcriptional repressors of the E(spl)/HES protein family (Bailey and Posakony 1995; Jarriault et al. 1995). As a result, achaete-scute (AS-C) proteins are prevented from converting this cell into a neuronal precursor, while factors determining the epidermal fate will either not be antagonized or may even be helped by the presence of NICD in the nucleus. In the neighboring cell, where NICD does not reach threshold levels in the nucleus, CSL protein will continue to repress its targets, Notch surface density will decline, and AS-C proteins will dominate. Among the transcriptional targets of AS-C proteins are the ligands for Notch receptors, in particular Delta, thus linking inactivation of Notch in one cell to activation of Notch in it neighbors.

This paradigm receives considerable, but mostly indirect, support. Biochemical evidence demonstrates that intramembranous proteolysis is induced by ligand in cell culture (Schroeter et al. 1998) and in vivo (Kidd et al. 1998). Genetic manipulation of the Notch protein in *Drosophila* also demonstrates that ligand-dependent intramembranous proteolysis occurs (Lecourtois and Schweisguth 1998; Struhl and Adachi 1998). However, these experiments do not demonstrate a requirement for Notch proteolysis in signal transduction. To address this deficiency we embarked on a reverse generate approach in mice. In our lab we have identified the precise peptide bond cleaved within the transmembrane domain (Gly1743 and Val 1744; Schroeter et al. 1998). A single amino acid substitution (V1744G) that impairs the processing efficacy by $\sim 90\%$ was introduced into the Notch 1 locus. Mice containing this point mutation are not viable and their phenotype resembles that of a Notch null (Huppert et al. 2000). Western blot analysis demonstrates that these embryos die while containing wild type levels of Notch1$^{V \Box G}$ protein. These results confirm that intramembranous proteolysis of Notch1 is indeed required for signal transduction. But how is Notch processed within the transmembrane domain, and how is this cleavage regulated?

Presenilin and Notch:
Trafficking or Regulated Intramembranous Proteolysis?

Several groups have realized the importance of presenilin proteins for intramembranous proteolysis of Notch1 (De Strooper et al. 1999; Ray et al. 1999a, b; Song et al. 1999; Struhl and Greenwald 1999; Ye and Fortini 1999; Ye et al. 1999). Most importantly, the phenotype of a double knockout of presenilin activity in both *C. elegans* (Levitan et al. 1996; Li and Greenwald 1997; Westlund et al. 1999) and mice (Donoviel et al. 1999; Herreman et al. 1999) bears a striking resemblance to the phenotype of complete loss of Notch signaling (Delapompa et al. 1997; Oka et al. 1996). However, one extremely important question still remains unsolved: are presenilin members of a new class of intramembranous enzymes (Wolfe et al. 1999a, b) or are they involved in trafficking (Naruse et al. 1998; Nishimura et al. 1999)? In support of the enzyme hypothesis, Mike Wolfe, Dennis Selkoe and their colleagues provided extremely intriguing evidence that conserved aspartyl residues in transmembrane domains 6 and 7 of presenilin 1 (PS1) are critical for biological activity (Steiner et al. 1999; Wolfe et al. 1999b); mutating D257 into either alanine, or a conservative change to Glutamate, forms a dominant negative inhibitor of presenilin activity. Notch cleavage is initiated at the cell surface in response to extracellular ligand binding (Kopan et al. 1996; Mumm et al. 2000), whereas Aβ is generated in multiple subcellular compartments in both the secretory and endocytic pathways. Furthermore, PS1 and PS2 are widely reported to be restricted to the endoplasmic reticulum (ER) and Golgi membranes (Haass 1997; Hardy 1997; Selkoe 1998, 1999). These observations are more consistent with PS1 having an indirect role in Notch and amyloid precursor protein (APP) metabolism, as suggested by studies reporting that the *C. elegans* Notch homologue LIN-12 had altered subcellular distribution in *sel-12* (a presenilin homologue) mutant animals (Levitan and Greenwald 1998) and that in PS1-deficient mammalian neurons the rate of APP secretion was abnormal (Naruse et al. 1998). Our demonstration that Notch forms complexes with presenilins in the secretory pathway was equally consistent with either hypothesis.

One way to directly differentiate between a role for presenilin in proteolysis or trafficking is simply to identify the step blocked by the aspartyl mutation in PS1. To focus our analysis on intramembranous proteolysis we transiently transfected human embryonic kidney (HEK) 293 cells with mNotch1ΔE (m1727v) (abbreviated ΔE), a truncated mouse Notch1 that undergoes PS1-regulated TM proteolysis constitutively in the absence of ligand and does not show alternative translation initiation (in the absence of the M1727V mutation, ΔE can be aberrantly translated from methionine 1727, located within the transmembrane domain, to generate a fragment that will translocate to the nucleus, thus mimicking the product of intracellular cleavage, "NICD," the Notch intracellular domain; Kopan et al. 1996). ΔE also contains six consecutive myc epitope tags at its C-terminus to facilitate detection of both the precursor and NICD following cleavage by γ-secretase, the enzymatic activity thought to be responsible for the RIP of Notch (De Strooper et al. 1999). Transient overexpression of wild-type PS1 had no effect on NICD levels compared to cells expressing endogenous PS1. However, when PS1 encoding an aspartate-to-alanine mutation at codon 257 (D257A PS1) was expressed. NICD levels were significantly reduced. Similar results were obtained when ΔE was transiently transfected into the stable CHO cell lines used to document the effect of this mutation on APP processing (data not shown; Wolfe et al. 1999b). Analysis of PS1 expression in these cells by immunoprecipitation followed by Western blot (IP-Western) revealed that overexpression of D257A PS1 had no effect on the accumulation of the N-terminal and C-terminal PS1 endoproteolytic cleavage fragments derived from endogenous PS1. Thus, D257A PS1 suppresses ΔE processing without altering the levels of endogenous PS1. This analysis also revealed that co-immunoprecipitation of presenilin proteins resulted in recovery of surface (biotinylated) Notch proteins, suggesting that both proteins can make their way together to the cell surface. These results are consistent with a role for presenilin proteins in Notch trafficking. To resolve this issue we compared the behavior of wild type presenilin and that of the D257A presenilin. If presenilin proteins function in trafficking of Notch to the cell surface, one would assume that the D257A presenilin proteins may be deficient in their ability to interact with Notch or fails to transport Notch to the cell surface (will not immunoprecipitate biotinylated Notch and may hold Notch hostage inside the cell). Alternatively, the aspartyl mutant presenilin may bind Notch in the ER and appear with it at the cell surface with similar kinetics but fail to facilitate Notch intramembranous proteolysis. Immunoprecipitation of presenilin proteins from cells co-transfected either with wild type or D257A presenilin proteins detects no difference in the amounts of Notch ΔE recovered, biotinylated surface protein or intracellular Notch. The only consistent difference seen in cells co-transfected with D257A presenilin is that Notch proteolysis is attenuated (Ray et al. 1999a). While these results do not support the hypothesis that presenilin proteins are involved in trafficking, they cannot differentiate between a role for presenilin proteins as enzymes or adapters. Collectively, these results raise the concern that inhibition of γ-secretase activity will lead to significant consequences within a short period of time, given that Notch signaling is required both for maintenance of hematopoietic stem cells (Varnum-Finney et al. 1998) and for the proper differentiation of T-cells within the thymus (Radtke et al. 1999).

The Regulation of Presenilin-Mediated Intramembranous Proteolysis of Notch by a Ligand-Induced Conformational Change

γ-secretase cleavage of APP protein occurs within multiple intracellular compartments, whereas Notch cleavage occurs past the trans-golgi, possibly at the cell surface. Since Notch and presenilin proteins associate in the ER, why isn't presenilin-mediated cleavage of Notch occurring constitutively, and how is the intramembranous proteolysis of Notch regulated by ligand binding?

Genetic analysis of gain of function Notch alleles has resulted in a hypothesis that the membrane proximal region, containing the LNR repeats and the two conserved cysteines, is negatively regulating in extra-membranous cleavage of Notch. To address the regulation of presenilin-dependent proteolysis, we adopted the general strategy of mimicking genetically defined activating mutations in the extracellular domain of Notch and comparing active and inactive Notch proteins at the biochemical level (Mumm et al. 2000). In one such active Notch molecule, two highly conserved cysteine residues (amino acids 1675 and 1682) have been mutated to serine (Lieber et al. 1993). To determine if proteolytic activation of Notch is inhibited in the presence of sequences other than Notch at the extracellular surface, we fused the extracellular domain of CD4 to $N^{\Delta E}$ (CD4·$N^{\Delta E}$). In both cases activity was gained; in addition, Western blot analysis revealed the appearance of a new proteolytic fragment migrating just below the uncleaved $N^{\Delta E}$ protein, consistent with an extracellular cleavage site (S2). Peptide sequencing shows that S2 cleavage occurs between Ala[1710] and Val[1711] residues, approximately 12 amino acids outside the transmembrane domain. This same peptide bond is cleaved by the metalloprotease TACE (*tumor necrosis factor α converting enzyme*; Brou et al. 2000). We have termed this product NEXT (for *Notch extracellular truncation*). NEXT is a naturally occurring equivalent of Notch ΔE-type constructs. This observation led to the hypothesis that the regulation of the intramembranous cleavage in Notch1 is accomplished by the presence of an inhibitory extracellular domain; ligand binding converts Notch from a protease-resistant conformation to a metalloprotease-sensitive one. The NEXT peptide is an intermediate; much like $N^{\Delta E}$, it is a substrate for the presenilin-dependent proteolytic apparatus. The proteolytic cascade hypothesis therefore postulates that NEXT is the precursor for NICD. If so, the hypothesis predicts that a block in NICD synthesis would result in the accumulation of NEXT, and vice versa: a block in NEXT production will eliminate the production of NICD. Most importantly, one also needs to demonstrate that ligand binding to Notch results in the induction of NEXT.

Three experiments have demonstrated that the first prediction holds true. First, point mutation substituting of V1744 to Leucine in $CD4 \cdot N^{\Delta E}$ fusion protein results in a reduction in NICD and accumulation of NEXT. Similarly, in PS1-deficient cells, the disappearance of NICD is accompanied by the accumulation of NEXT. More dramatically, repeated treatment of these Notch proteins with γ-secretase inhibitors lead to a substantial increase in the NEXT protein. The second prediction was tested using a metalloprotease inhibitor; we demonstrated that blocking NEXT proteolysis does not interfere with NICD production from $N^{\Delta E}$ but results in the loss of NEXT and NICD from activated Notch proteins containing an extracellular domain. The last prediction of the proteolytic cascade model, that ligand binding induces NEXT proteolysis, was verified by co-culturing Notch transfected cells with ligand transfected cells. Collectively, these data demonstrate that Notch is a dual address protein, activated by a ligand-induced "ectodomain shedding" like event (S2 cleavage) that generates the intermediate NEXT (Mumm et al. 2000; Brou et al. 2000). NEXT is converted to NICD by a presenilin-dependent proteolysis. This sequence of events is reminiscent of APP proteolysis: γ-secretase cleavage occurs only after β-secretase cleavage.

Presenilin Proteins:
A Central Component in a New Signaling Paradigm

The RIP paradigm suggests that multiple dual address proteins are transported to various cellular locations where the protein awaits a triggering event. This event (i.e., ligand binding) will lead to the release a domain harboring a second "zip code;" this domain will then translocate to a new location where it has a biological function. This paradigm was first described in bacteria: the $sigma^E$ and $sigma^K$ proteins are transcription factors regulating sporulation that are activated upon proteolytic cleavage of their "pro" form (Kroos et al. 1989; LaBell et al. 1987; Lu et al. 1990, 1995). Under starvation conditions, which induce sporogenesis, pro-σ^E and pro-σ^K are translocated to the cell membrane where they are proteolytically cleaved, presumably within the membrane, and released to bind chromosomal targets (Brown et al. 2000; Rudner et al. 1999). The bacterial pheromone cAD1 is an octapeptide generated from a larger type II transmembrane precursor via a proteolytic cascade similar to the one described above for Notch. The sterol response element binding protein (SREBP) is also a dual address type II transmembrane protein released by RIP (Brown et al. 2000). Proteolysis of all three proteins requires the activity of related, multi-transmembrane domain proteins containing two conserved amino acid consensus sequences – HxxGH in one transmembrane domain and LDG in apposition to it in another – capable of Zn++ coordination. The prevailing hypothesis is that they represent a novel family of intramembranous metalloproteases. Similarly, the presenilin proteins are conserved, multi-transmembrane domain proteins suspected of acting as aspartyl proteases. Presenilin activity is required downstream of an extracellular cleavage for RIP of APP and Notch. Recently, a third substrate for presenilin activity was described: Ire1, the human homolog of a yeast regulator of the unfolded protein response (UPR). Ire1 is an ER resident protein that, in response to the accumulation of unfolded proteins, undergoes a presenilin-dependent cleavage and translocates to the nucleus where it acts as a chaperone protein.

Conclusions and Future Directions

The presenilin protein, as its name implies, was cloned because of its involvement in the onset of Alzheimer's disease and, independently, its involvement in Notch signaling. These two fields have now merged with others to suggest a role for presenilin proteins as members of a novel class of enzymes, capable of cleaving within the transmembrane domain. Continued effort from developmental and cell biologists, human geneticists and clinicians will undoubtedly provide better insight into the role of presenilins and APP in a healthy brain and body. The involvement of presenilin proteins in Notch signaling and unfolded protein response may limit the use of first generation γ-secretase inhibitors. However, as the final number of biological processes that are dependent on presenilin becomes known, it will facilitate identification of the means by which the rate of Alzheimer's disease onset could be slowed down without deleterious side effects.

References

Angerer LM, Angerer RC (1999) Regulative development of the sea urchin embryo: Signalling cascades and morphogen gradients [Review]. Sem Cell Dev Biol 10: 327–334

Artavanis-Tsakonas S, Simpson P (1991 Choosing a cell fate: a view from the Notch locus. [Review]. Trends Genet 7: 403–408

Artavanis-Tsakonas S, Rand MD, Lake RJ (1999) Notch signaling: cell fate control and signal integration in development. [Review] Science 284: 770–776

Bailey AM, Posakony JW (1995) Suppressor of Hairless directly activates transcription of Enhancer of split complex genes in response to Notch receptor activity. Genes Dev 9: 2609–2622

Blaumueller CM, Qi HL, Zagouras P, Artavanis-Tsakonas S (1997) Intracellular cleavage of Notch leads to a heterodimeric receptor on the plasma membrane. Cell 90: 281–291

Bray S (1998) Notch signalling in Drosophila: three ways to use a pathway. Sem Cell Dev Biol 9: 591–597

Brou C, Logeat F, Gupt N, Bessia C, LeBail O, Doedens JR, Cumano A, Roux P, Black RA, Israel A (2000) A novel proteolytic cleavage involved in Notch signaling: the role of the disintegrin-metalloprotease TACE. Molecular Cell 5: 207–216

Brown MS, Ye J, Rawson RB, Goldstein JL (2000) Regulated intramembrane proteolysis: a control mechanism conserved from bacteria to humans includes the SREBPs (sterol regulatory element-binding proteins), transmembrane proteins of the ER whose cytosolic (Review). Cell 100: 391–398

Collier JR, Monk NAM, Maini PK, Lewis JH (1996) Pattern formation by lateral inhibition with feed-back – a mathematical model of Delta-Notch intercellular signalling. J Theoret Biol 183: 429–446

Delapompa J, Wakeham A, Correia KM, Samper E, Brown S, Aguilera RJ, Nakano T, Honjo T, Mak TW, Rossant J, Conlon RA (1997) Conservation of the Notch signalling pathway in mammalian neurogenesis. Development 124: 1139–1148

De Strooper B, Annaert W, Cupers P, Saftig P, Craessaerts K, Mumm JS, Schroeter EH, Schrijvers V, Wolfe MS, Ray WJ, Goate A, Kopan R (1999) A presenilin-1-dependent gamma-secretase-like protease mediates release of Notch intracellular domain. Nature 398: 518–522

Donoviel DB, Hadjantonakis AK, Ikeda M, Zheng H, Hyslop PS, Bernstein A (1999) Mice lacking both presenilin genes exhibit early embryonic patterning defects. Genes Dev 13: 2801–2810

Greenwald I (1994) Structure/function studies of lin-12/Notch proteins [Review]. Curr Opin Genet Dev 4: 556–562

Greenwald I (1998) Lin-12/Notch signaling – lessons from worms and flies. Genes Dev 12: 1751–1762

Haass RN (1997) Sanctioning madness. Foreign Affairs 76: 74–85

Hardy J (1997) The amyloid cascade hypothesis of AD – decoy or real McCoy – Reply. Trends Neurosci 20: 559

Heitzler P, Simpson P (1991) The choice of cell fate in the epidermis of Drosophila. Cell 64: 1083–1092

Heitzler P, Simpson P (1993) Altered epidermal growth factor-like sequences provide evidence for a role of Notch as a receptor in cell fate decisions. Development 117: 1113–1123

Herreman A, Hartmann D, Annaert W, Saftig P, Craessaerts K, Serneels L, Umans L, Schrijvers V, Checler F, Vanderstichele H, Baekelandt V, Dressel R, Cupers P, Huylebroeck D, Zwijsen A, Van Leuven F, De Strooper B (1999) Presenilin 2 deficiency causes a mild pulmonary phenotype and no changes in amyloid precursor protein processing but enhances the embryonic lethal phenotype of presenilin 1 deficiency. Proc Natl Acad Sci USA 96: 11872–11877

Hsieh JJD, Hayward SD (1995) Masking of the CBF1/RBPjϰ transcriptional repression domain by Epstein-Barr virus EBNA2. Science 268: 560–563

Hsieh JJ, Henkel T, Salmon P, Robey E, Peterson MG, Hayward SD (1996) Truncated mammalian Notch1 activates CBF1/RBPJk-repressed genes by a mechanism resembling that of Epstein-Barr virus EBNA2. Mol Cell Biol 16: 952–959

Hsieh JJD, Zhou SF, Chen L, Young DB, Hayward SD (1999) CIR, a corepressor linking the DNA binding factor CBF1 to the histone deacetylase complex. Proc Natl Acad Sci USA 96: 23–28

Huppert SS, Le A, Schroeter EH, Mumm JS, Saxena MT, Milner LA, Kopan R (2000) Embryonic lethality in mice homozygous for a processing-deficient allele of Notch1. Nature 405: 966–970

Jarriault S, Brou C, Logeat F, Schroeter EH, Kopan R, Israel A (1995) Signalling downstream of activated mammalian Notch. Nature 377: 355–358

Joutel A, Tournier-Lasserve E (1998) Notch signalling pathway and human diseases [Review]. Sem Cell Dev 9: 619–625

Kao HY, Ordentlich P, Koyanonakagawa N, Tang Z, Downes M, Kintner CR, Evans RM, Kadesch T (1998) A histone deacetylase corepressor complex regulates the Notch signal transduction pathway. Genes Dev 12: 2269–2277

Kidd S, Lieber T, Young MW (1998) Ligand-induced cleavage and regulation of nuclear entry of Notch in Drosophila melanogaster embryos. Genes Dev 12: 3728–3740

Kimble J, Henderson S, Crittenden S (1998) Notch/Lin-12 signaling – transduction by regulated protein slicing. Trends Biochem Sci 23: 353–357

Kopan R, Turner DL (1996) The Notch pathway: democracy and aristocracy in the selection of cell fate. Curr Opin Neurobiol 6: 594–601

Kopan R, Schroeter EH, Weintraub H, Nye JS (1996) Signal transduction by activated mNotch: Importance of proteolytic processing and its regulation by the extracellular domain. Proc Natl Acad Sci USA 93: 1683–1687

Kroos L, Kunkel B, Losick R (1989) Switch protein alters specificity of RNA polymerase containing a compartment-specific sigma factor. Science 243: 526–529

LaBell TL, Trempy JE, Haldenwang WG (1987) Sporulation-specific sigma factor sigma 29 of Bacillus subtilis is synthesized from a precursor protein, P31. Proc Natl Acad Sci USA 84: 1784–1788

Lecourtois M, Schweisguth F (1998) Indirect evidence for Delta-dependent intracellular processing of Notch in Drosophila embryos. Curr Biol 8: 771–774

Levitan D, Greenwald I (1998) Effects of SEL-12 presenilin on LIN-12 localization and function in Caenorhabditis elegans. Development 125: 3599–3606

Levitan D, Doyle TG, Brousseau D, Lee MK, Thinakaran G, Slunt HH, Sisodia SS, Greenwald I (1996) Assessment of normal and mutant human presenilin function in Caenorhabditis elegans. Proc Natl Acad Sci USA 93: 14940–14944

Lewis J (1996) Neurogenic genes and vertebrate neurogenesis. Curr Opin Neurobiol 6: 3–10

Lewis J (1998a) Notch signalling – a short cut to the nucleus. Nature 393: 304–305

Lewis J (1998b) Notch signalling and the control of cell fate choices in vertebrates. Sem Cell Dev Biol 9: 583–589

Li XJ, Greenwald I (1997) Hop-1, a Caenorhabditis elegans presenilin, appears to be functionally redundant with Sel-12 presenilin and to facilitate Lin-12 and Glp-1 signaling. Proc Natl Acad Sci USA 94: 12204–12209

Lieber T, Kidd S, Alcamo E, Corbin V, Young MW (1993) Antineurogenic phenotypes induced by truncated Notch proteins indicate a role in signal transduction and may point to a novel function for Notch in nuclei. Genes Dev 7: 1949–1965

Logeat F, Bessia C, Brou C, Lebail O, Jarriault S. Seidah NG, Israel A (1998) The Notch1 receptor is cleaved constitutively by a furin-like convertase. Proc Natl Acad Sci USA 95: 8108–8112

Lu S, Halberg R, Kroos L (1990) Processing of the mother-cell sigma factor, sigma K, may depend on events occurring in the forespore during Bacillus subtilis development. Proc Natl Acad Sci USA 87: 9722–9726

Lu S, Cutting S, Kroos L (1995) Sporulation protein SpoIVFB from Bacillus subtilis enhances processing of the sigma facto precursor Pro-sigma K in the absence of other sporulation gene products. J Bacteriol 177: 1082–1085

Miele L, Osborne B (1999) Arbiter of differentiation and death: Notch signaling meets apoptosis [Review]. J Cell Physiol 181: 393–409

Morel V, Schweisguth F (2000) Repression by suppressor of Hairless and activation by Notch are required to define a single row of single-minded expressing cells in the Drosophila embryo. Genes Dev 14: 377–388

Mumm JS, Schroeter EH, Saxena MT, Griesemer A, Tian X, Pan DJ, Ray WJ, Kopan R (2000) A ligand-induced extracellular cleavage regulates γ-secretase-like proteolytic activation of Notch1. Mol Cell 5: 197–206

Naruse S, Thinakaran G, Luo JJ, Kusiak JW, Tomita T, Iwatsubo T, Qian XZ, Ginty DD, Price DL, Borchelt DR, Wong PC, Sisodia SS (1998) Effects of PS1 deficiency on membrane protein trafficking in neurons. Neuron 21: 1213–1221

Nishimura M, Yu G, Levesque G, Zhang DM, Ruel L, Chen F, Milman P, Holmes E, Liang Y, Kawarai T, Jo E, Supala A, Rogaeva E, Xu DM, Janus C, Levesque L, Bi Q, Duthie M, Rozmahel R, Mattila K, Lannfelt L, Westaway D, Mount HTJ, Woodgett J, Fraser PE, St George-Hyslop P (1999) Presenilin mutations associated with Alzheimer disease cause defective intracellular trafficking of beta-catenin, a component of the presenilin protein complex. Nature Med 5: 164–169

Oka C, Nakano T, Wakeham A, Delapompa JL, Mori C, Sakai T, Okazaki S, Kawaichi M, Shiota K, Mak TW, Honjo T (1996) Disruption of the mouse Rbp-J(Kappa) gene results in early embryonic death (Vol 121, Pg 3291, 1995). Development 122: 405–407

Olave I, Reinberg D, Vales LD (1998) The mammalian transcriptional repressor Rbp (Cbf1) targets Tfiid and Tfiia to prevent activated transcription. Genes Dev 12: 1621–1637

Posakony JW (1994) Nature versus nurture: asymmetric cell divisions in *Drosophila* bristle develop-
    ment [comment]. [Review]. Cell 76: 415–418

Radtke F, Wilson A, Stark G, Bauer M, van Meerwijk J, MacDonald HR, Aguet M (1999) Deficient T cell
    fate specification in mice with an induced inactivation of Notch1. Immunity 10: 547–558

Rand DM, Grimm MLM, Artavanis-Tsakonas S, Patriub V, Blacklow CS, Sklar CJ, Aster CJ (2000). Cal-
    cium depletion dissociates and activates heterodimeric Notch receptors. Mol Cell Biol 20

Ray WJ, Yao M, Mumm J, Schroeter EH, Saftig P, Wolfe M, Selkoe DJ, Kopan R, Goate AM (1999a) Cell
    surface presenilin-1 participates in the gamma-secretase-like proteolysis of notch. J Biol Chem 274:
    36801–36807

Ray WJ, Yao M, Nowotny P, Mumm J, Zhang WJ, Wu JY, Kopan R, Goate AM (1999b) Evidence for a
    physical interaction between presenilin and Notch. Proc Natl Acad Sci USA 96: 3263–3268

Robey E (1999) Regulation of T Cell fate by Notch [Review]. Ann Rev Immunol 17: 283–295

Rooke JE, Xu T (1998) Positive and negative signals between interacting cells for establishing neural
    fate [Review]. Bioessays 20: 209–214

Rudner DZ, Fawcet P, Losick R (1999) A family of membrane-embedded metalloproteases involved in
    regulated proteolysis of membrane-associated transcription factor. Proc Ntl Acad Sci USA 96:
    14765–14770

Saito T, Watanabe N (1998) Positive and negative thymocyte selection. Crit Rev Immunol 18: 359–370

Schroeter EH, Kisslinger JA, Kopan R (1998) Notch-1 signalling requires ligand-induced proteolytic
    release of intracellular domain [see comments]. Nature 393: 382–386

Selkoe DJ (1998) The cell biology of beta-amyloid precursor protein and presenilin in Alzheimer's dis-
    ease. Trends Cell Biol 8: 447–453

Selkoe DJ (1999) Translating cell biology into therapeutic advances in Alzheimer's disease [Review].
    Nature 399: A23–A31

Song WH, Nadeau P, Yuan ML, Yang XD, Shen J, Yankner BA (1999) Proteolytic release and nuclear
    translocation of Notch-1 are induced by presenilin-1 and impaired by pathogenic presenilin-1 muta-
    tions. Proc Natl Acad Sci USA 96: 6959–6963

Steiner H, Duff K, Capell A, Romig H, Grim MG, Lincoln S, Hardy J, Yu X, Picciano M, Fechteler K, Cit-
    ron M, Kopan R, Pesold B, Keck S, Baader M, Tomita T, Iwatsubo T, Baumeister R, Haass C (1999) A
    loss of function mutation of presenilin-2 interferes with amyloid beta-peptide production and notch
    signaling. J Biol Chem 274: 28669–28673

Struhl G, Adachi A (1998) Nuclear access and action of Notch in vivo. Cell 93: 649–660

Struhl G, Greenwald I (1999) Presenilin is required for activity and nuclear access of Notch in *Drosoph-
    ila*. Nature 398: 522–525

Varnum-Finney B, Purton LE, Yu M, Brashemstein C, Flowers D, Staats S, Moore KA, Leroux I,
    Mann R, Gray G, Artavanis-Tsakonas S, Bernstein ID (1998) The Notch ligand, Jagged-1, influences
    the development of primitive hematopoietic precursor cells. Blood 91: 4084–4091

Westlund B, Parry D, Clover R, Basson M, Johnson CD (1999) Reverse genetic analysis of *Caenorhabdi-
    tis elegans* presenilins reveals redundant but unequal roles for sel-12 and hop-1 in Notch-pathway
    signaling. Proc Natl Acad Sci USA 96: 2497–2502

Wolfe MS, De los Angeles J, Miller DD, Xia WM, Selkoe DJ (1999a) Are presenilins intramembrane-
    cleaving proteases? Implications for the molecular mechanism of Alzheimer's disease. Biochemistry
    38: 11223–11230

Wolfe MS, Xia WM, Ostaszewski BL, Diehl TS, Kimberley WT, Selkoe DJ (1999b) Two transmembrane
    aspartates in presenilin-1 required for presenilin endoproteolysis and gamma-secretase activity.
    Nature 398: 513–517

Wu JY, Rao Y (1999) Fringe: defining borders by regulating the Notch pathway. Curr Opin Neurobiol 9:
    537–543

Ye YH, Fortini ME (1999) Apoptotic activities of wild-type and Alzheimer's disease-related mutant pre-
    senilins in *Drosopila melanogaster*. J Cell Biol 146: 1351–1364

Ye YH, Lukinova N, Fortini ME (1999) Neurogenic phenotypes and altered Notch processing in *Dro-
    sophila* presenilin mutants. Nature 398: 525–529

The Putative Role of Presenilins in the Transmembrane Domain Cleavage of Amyloid Precursor Protein and Other Integral Membrane Proteins

B. De Strooper, A. Herreman, P. Cupers, K. Craessaerts, L. Serneels, and W. Annaert

Summary and Introduction

Missense mutations in the presenilin (PS)-1 and -2 genes are major causes of familial Alzheimer's disease (AD), acting apparently in a dominant fashion (Levy-Lahad et al. 1995; Rogaev et al. 1995; Sherrington et al. 1995). Although the exact pathogenic mechanism underlying the disease process remains to be further elucidated, it is fairly established that almost all PS missense mutations affect the processing of the amyloid precursor protein (APP). The result is an increased secretion of the longer form of the amyloid peptide (Borchelt et al. 1996; Duff et al. 1996; Lemere et al. 1996; Tomita et al. 1997; Xia et al. 1997). This peptide constitutes the major component of the amyloid plaques in patients. Interestingly several of the PS mutations appear also to enhance the sensitivity of cells, and in particular neurons, to apoptotic stimuli (Deng et al. 1996; Guo et al. 1997; Janicki and Monteiro 1997; Vito et al. 1997; Wolozin et al. 1996). In principle both mechanisms could contribute to the pathogenesis of AD (Zhang et al. 1998). Insight into the biological functions of the PS in the cell is probably not only important for understanding the pathogenesis of the familial form of AD, but also of the sporadic form of AD. Interestingly, recent findings suggest that PS is involved in the proteolysis of the transmembrane domains of APP, Notch, APLP-1, and possibly Ire-1, and could therefore function as a molecular switch linking proteolysis to intracullular signaling (Annaert and De Strooper 1999).

The Presenilins

Presenilin (PS)-1 and its homologue, PS-2, consist of 467 and 448 aminoacid residues, respectively (Fig. 1). They are highly hydrophobic and span the membranes of the endoplasmic reticulum (ER) most likely eight times (although alternative models do exist). At least the hydrophilic N-terminus and a relatively large loop domain between hydrophobic domains 6 and 7 are oriented to the cytoplasm (De Strooper et al. 1997; Doan et al. 1996; Lehmann et al. 1997; Li and Greenwald, 1996, 1998). These domains are therefore available for interactions with cytoplasmic proteins such as the catenins (Murayama et al. 1998; Tesco et al. 1998; Yu et al. 1998; Zhang et al. 1998; Zhou et al. 1997), among others. Shortly after biosynthesis, the PS are cleaved in their hydrophilic loop domain and a complex of

Research and Perspectives in Alzheimer's Diseases
Beyreuther/Christen/Masters (Eds.)
Neurodegenerative Disorders
© Springer-Verlag Berlin Heidelberg 2001

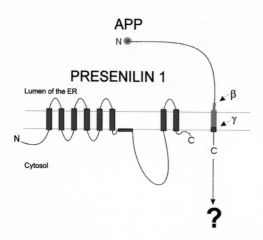

Fig. 1. Amyloid Precursor Protein (APP) and PS. PS is schematically drawn as an eight-transmembrane domain containing protein, with the aminoterminal (N), the loop- and the carboxyterminal (C) domains oriented to the cytoplasm. It should be noticed that PS is cleaved in the loop domain by an unknown protease, and that the resulting fragments constitute the active complex. APP is also displayed. The β-secretase (β) and the γ-secretase (γ) sites are indicated. The red box between these two cleavage sites is the amyloid peptide. ER, endoplasmic reticulum; SP, signal peptide

a ~30 kDa N-terminal fragment (NTF) and a ~20 kDa C-terminal fragment (CTF) is formed. Both fragments exist as heterodimeric complexes and are part of a larger complex of about 250 kDa that, in addition, harbors β-catenin among other proteins (Levesque et al. 1999; Yu et al. 1998; Capell et al. 1998; Thinakaran et al. 1996; Yu et al. 1998). Earlier studies that used overexpression of the genes encoding PS-1 and PS-2 suggested a localization in the ER and in the Golgi apparatus (De Strooper et al. 1997; Kovacs et al. 1996; Walter et al. 1996). Recent electron microscopy and confocal microscopy studies combined with subcellular fractionation studies, indicate that the subcellular distribution of endogenously synthesized PS in neurons is indeed mainly limited to the ER, the intermediate compartment and an early *cis*-Golgi compartment (Annaert et al. 1999; Capell et al. 1997; Culvenor et al. 1997; Lah et al. 1997). However, as has been suggested in some recent publications, it cannot be excluded that a minor amount of PS is expressed close to or at the cell surface (Georgakopoulos et al. 1999; Ray et al. 1999).

Presenilins Control Proteolytic Processing of APP

Shortly after the discovery that missense mutations of the PS were a major cause of familial AD (Sherrington et al. 1995), a link was made with the proteolytic processing of APP and the generation of the amyloid peptide (Citron et al. 1997; Duff et al. 1996; Scheuner et al. 1996). APP is a type-I integral membrane protein and is cleaved by a series of enzymes, called α- and β-secretase, which results in the shedding of its ectodomain and the production of C-terminal APP-"stubs," named α-stub and β-stub, respectively (Haass and Selkoe 1993). These membrane-bound stubs are the substrates for γ-secretase(s), which result in the release of the amyloid peptide from the β-stubs and the p3-peptide from the α-stubs. Direct proof for the role of PS-1 in γ-secretase processing of APP has come from studies using PS-1-deficient neurons (De Strooper et al. 1998) and cells overexpressing dominant negative mutant forms of PS-1 and -2 (Wolfe et al. 1999c; Steiner et al. 1999). In those cells a strong inhibition of the amyloid- and p3-peptide secretion was observed. A concomitant accumulation of the direct precursors of these peptides, that is, the α-and β-secretase generated C-terminal APPstubs, demonstrated that PS-1 deficiency resulted in inhibition of the normal γ-secretase processing of APP. Also the turnover of the C-terminal membrane-bound fragments of APP-like-protein-1 (APLP1), a homologue of APP, is decreased in PS-1-deficient neurons (Naruse et al. 1998).

Presenilins and Notch Signalling

Notch processing is also dependent on PS. The link between Notch function and PS was originally made in *C. elegans* (Baumeister et al. 1997; Levitan and Green-wald 1995). PS-1 knockout mice furthermore show some signs of deficient Notch signalling (Hartmann et al. 1999; Shen et al. 1997; Wong et al. 1997). It should, however, be noticed that the phenotype of the PS1-deficient mice differs in several important aspects from the previously described Notch-1-deficient phenotype (Table 1). For instance PS-1-/-embryos suffer from severe brain hemorrhage and are viable until birth, whereas Notch-1-deficient animals do not even survive E11 (Conlon et al. 1995; Swiatek et al. 1994; see Table 1). Only when the PS-2 gene is also inactivated does a full Notch-deficient phenotype become obvious in mice (Herreman et al. 1999; Donoviel et al. 1999).

Notch-1 is a large, type 1 integral membrane protein involved in complex cell fate decisions (Artavanis-Tsakonas et al., 1999). Following ligand binding the cytoplasmic domain of the mature Notch protein becomes released by a proteolytic cleavage step in or close to the transmembrane domain (Fig. 2; Schroeter et al. 1998). The cytoplasmic domain is a signalling factor, which is assumed to be translocated in complex with members of the CSL [CBF1, Su(H), Lag-1] family of DNA binding proteins to the nucleus. This brings genes involved in myogenesis, neuronal differentiation, and hematopoiesis under the control of Notch (Artavanis-Tsakonas et al. 1999). The deficient Notch signalling (De Strooper et al. 1999; Donoviel et al. 1999; Herreman et al. 1999; Song et al. 1999) and the

Table 1. Comparison of selected phenotypical features of the PS-1 single – deficient, the PS-1 and 2 double – deficient, and the Notch-1 – deficient mice.

	PS-1-/-	PS-1-/-PS-2-/-	Notch 1-/-
• Lethality	Late	Early	Early
• Posterior development	Truncated	Truncated	Truncated
• Somitogenesis	Disturbed	Severely disturbed	Severely disturbed
• Heart	Normal	Distended pericardia	Distended pericardia
• Brain	Underdeveloped	Underdeveloped	Underdeveloped
• Yolk sac (9.5d)	Normal	RBC in vascular plexus; no vessels	Normal

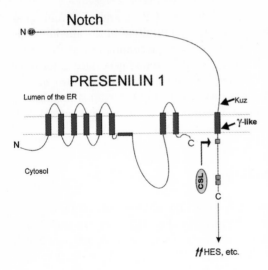

Fig. 2. Notch processing is dependent on PS expression. Notch is a large type I integral membrane protein involved in cell fate decision. After binding to its ligand, Notch is processed in its ectodomain and in or close to the transmembrane domain releasing the intracellular domain. The intracellular domain interacts with DNA binding proteins of the CSL family and is translocated to the nucleus, where gene expression is induced. The second cleavage is dependent on the presence of PS. ER, endoplasmic reticulum; γ-like, γ-secretase-like site; SP, signal peptide

classical Notch phenotypes in PS-deficient *Drosophila* (Struhl and Greenwald 1999; Ye et al. 1999) and *C. elegans* (Baumeister et al., 1997; Levitan and Greenwald, 1995) are now explained at the molecular level by the need for PS to cleave the cytoplasmic domain of Notch (De Strooper et al. 1999; Song et al. 1999; Struhl and Greenwald 1999). The finding that a γ-secretase-like activity is involved in Notch processing was further corroborated by the fact that established γ-secretase inhibitors also inhibit Notch intramembranous processing (De Strooper et al. 1999).

Presenilins and the Unfolded Protein Response Signalling Pathway

Cells that express clinical mutants of PS not only show abnormal processing of APP, but also display disturbed stress responses and apoptosis when treated with the glycosylation inhibitor tunicamycin or with the calcium ionophore A23187 (Katayama et al. 1999). At the molecular level, the induction of a series of protein folding chaperones like GRP78/Bip, GRP94 and PDI (protein disulphide isomerase) is compromised in the cells with the mutant PS (Katayama et al. 1999) and also in cells derived from PS-1-deficient mice (Niwa et al. 1999).

The expression of these chaperones is controlled by the *u*nfolded *p*rotein *r*esponse (UPR) signalling pathway. One important upstream effector of the UPR pathway is Ire1p, a type I integral membrane protein like APP and Notch (Fig. 3). Its aminoterminal domain is oriented to the lumen of the ER and is believed to act as a detector of unfolded proteins. In response to the accumulation of unfolded proteins in the ER (e.g., after tunicamycin treatment), Ire1p oligomerizes and becomes (auto)phosphorylated. The cytoplasmic domain of Ire1p con-

Fig. 3. Ire1p and the stress response. Ire1p is a type I integral membrane protein of the endoplasmic reticulum (ER). Its aminoterminal domain is putatively sensing the presence of unfolded proteins. This signal is transmitted to the cytoplasmic domain, which is thought to be proteolytically released and to be transported to the nucleus, where it is proposed to be involved in the splicing of a control element in mRNA coding for Hac-like transcription factors involved in the unfolded stress response, at least in yeast. The model remains speculative, but explains the available data (see text). UPR, unfolded protein response; γ, γ-secretase site; SP, signal peptide

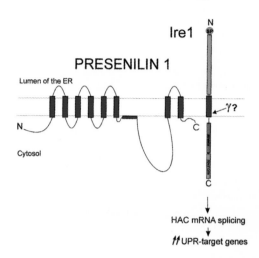

tains a serine/threonine kinase domain and an endoribonuclease domain in the carboxyterminal part (Fig. 3). This domain is, at least in yeast, involved in a unique splicing event, the excision of an inhibitory 252 nucleotide intron from the Hac1p mRNA, allowing translation of the mRNA into active Hac1p protein. The Hac1p protein is a transcription factor that binds to the "UPR element" in the promoters of the UPR proteins and therefore controls the transcription of the chaperones involved in the UPR. While the mammalian homologue of Hac1p has not been identified yet, mammalian Ire1p is perfectly able to splice the yeast Hac1p mRNA. It is therefore likely that mammalian Ire1p operates in a similar fashion as its yeast counterpart.

Niwa et al. (1999) demonstrated that during UPR the cytoplasmic part of Ire1p is translocated to the nucleus. This finding could suggest that this domain is specifically cleaved in response to the UPR induction. In fibroblasts derived from PS-1-deficient mice the nuclear translocation of the cytoplasmic domain of Ire1p in response to the induction of the UPR appeared to be strongly impaired (Niwa et al. 1999). Given the role of PS-1 in the processing of Notch and APP, the authors proposed that PS could also be needed for the putative proteolysis of Ire1p. However, several questions remain to be answered. First, direct evidence for the proteolysis of Ire1p and the effect of PS was not provided, and the proposed involvement of proteolysis in this signal transduction pathway therefore remains largely speculative. Furthermore, a mammalian homologue of Hac1p remains to be identified. Finally, it is unclear to what extent the PS deficiency in transgenic mice affects the UPR in vivo. One would predict that a complete inactivation of this pathway would compromise very early steps in embryonic development. PS-1 knockout mice survive beyond embryonic day 16 (Hartmann et al. 1999; Shen et al. 1997; Wong et al. 1997), PS-2 knockout mice are vital (Herreman et al. 1999; Steiner et al. 1999) and PS-1/PS-2 double knockout mice (Donoviel et al. 1999; Herreman et al. 1999) survive until embryonic day 9 (Table 1). In addition, the major phenotypical alterations can be explained by deficient Notch signalling in these mice, precluding the need to invoke an early impairment of the UPR pathway to explain the phenotype. Further studies are urgently needed to establish the relative contribution of each signalling pathway in the complex developmental disorder observed in the PS double-deficient embryos.

SREBP and SCAP Processing and Signalling

While the concept that APP is cleaved by proteases in its transmembrane domain was already proposed some time ago, it remains a provocative idea, since water is needed to hydrolyze peptide bonds. However, although the molecular mechanisms involved remain unclear, proteolysis in or close to the transmembrane domain of integral membrane proteins and release of the cytoplasmic domain appear to be pivotal in several signalling pathways. The best-studied example is the regulation of cholesterol homeostasis by SREBP/SCAPP (Brown and Goldstein 1997). *Sterol Regulatory Element Binding Protein* (SREBP) is a hairpin-like, membrane bound protein. In sterol-deprived cells, a site 1 protease (S1P) of the subtilisin family of proteases, cleaves SREBP in its luminal loop domain (Sakai et al. 1998). A consecutive cleavage by a hydrophobic site 2 zinc metalloprotease (S2P, Rawson et al. 1997) in the transmembrane domain occurs then by default, releasing the cytoplasmic aminoterminal domain of SREBP (Fig. 4). This domain is a DNA binding protein that translocates to the nucleus to activate transcription of genes like 3-hydroxy-3-methylglutaryl co-

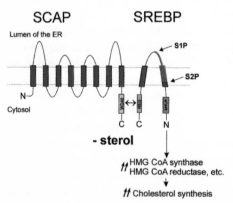

Fig. 4. Regulation of cholesterol biosynthesis by SREBP. Sterol regulatory element binding protein (SREBP) is a hairpin-like transcription factor residing in the endoplasmic reticulum (ER). Under cholesterol-deficient conditions, SREBP is cleaved first by the site 1 protease (S1P) in the luminal domain and then by the site 2 protease (S2P) in the transmembrane domain, releasing the aminoterminal cytoplasmic domain, which then migrates to the nucleus and controls transcription of genes involved in cholesterol biosynthesis. A subtilisin type of protease that resides in the *cis*-Golgi performs the site 1 cleavage. SCAP (SREBP cleavage activating protein) is responsible for the transport of SREBP to the compartment where the S1P is residing. CoA, coenzyme A

enzyme A(HMG CoA) synthase, HMG CoA reductase and other enzymes that are involved in cholesterol biosynthesis.

A cholesterol sensing protein, called SREBP cleavage activating protein (SCAP) regulates the first cleavage by S1P. SCAP is a multitransmembrane domain containing protein that resides in the ER, like SREBP. Upon lowering cholesterol, SCAP escorts SREBP from the ER to a post-ER compartment where S1P is operating (DeBose-Boyd et al. 1999; Nohturfft et al. 1999). SCAP thus appears to be what could be called a "transport chaperone." The proposed mechanism therefore couples vesicular transport to proteolysis and signal transduction (DeBose-Boyd et al. 1999).

It should be pointed out from the beginning that none of the proteins involved in SREBP processing is likely to be involved in APP processing (De Strooper et al. 1999; Ross et al. 1998). However, some analogies have been drawn between SREBP/SCAP and APP/PS (Brown and Goldstein 1997; Chan and Jan, 1999; De Strooper et al. 1998, 1999). PS contain eight transmembrane domains like SCAP, and are located in the same subcellular compartments as SCAP (Annaert et al. 1999). On the other hand, the S2P protease that cleaves SREBP in the transmembrane domain is a highly hydrophobic protein, like PS, so the analogy could be made there as well. The SREBP/SCAP model is therefore used by all participants in the debate on the exact role of PS in the γ-secretase processing of APP. One hypothesis suggests that PS themselves could be aspartyl proteases responsible for γ-secretase cleavage of APP (Wolfe et al. 1999a). This hypothesis is based on the observation that substitution of either one of the two critical aspartate residues in transmembrane domain 6 and 7 of PS by other amino acid residues results in a "dominant negative" effect on APP processing. This effect is very similar to the effects on APP processing in cells containing inactivated PS-1 genes (De Strooper et al. 1998; Steiner et al., 1999; Wolfe et al. 1999c). APP processing can indeed be inhibited using aspartyl protease inhibitors (Wolfe et al. 1999b), and the aspartyl residues in the PS have been conserved in evolution, suggesting that they are involved in an important function (Brockhaus et al. 1998; Leimer et al. 1999).

While the "PS is γ-secretase" hypothesis is without any doubt intellectually attractive, caution is required. First, direct proof that PS are proteases is lacking and the interpretation that the aspartates are part of a catalytic domain is still speculative. It could well be that the dominant negative effects observed with the aspartate mutations are consequences of structural alterations of the PS interfering with for instance binding to a yet unknown protein. Furthermore the subcellular localization of the PS and the compartments where γ-secretase activity is believed to occur do not overlap completely (Annaert et al. 1999). This "spatial paradox" remains to be explained (Annaert and De Strooper 1999). Finally, refined analysis of the specificity of γ-secretase processing using APP mutants suggests that it is unlikely that only one protease is involved in this process (Murphy et al. 1999). The alternative interpretation then is that the PS are only important cofactors in the cleavage of transmembrane domain proteins like APP (De Strooper et al. 1998), APLP-1 (Naruse et al. 1998) and Notch (De Strooper et al.

1999). This could theoretically be the creation of a microdomain environment in the membrane that, for instance, allows the necessary water molecules to be brought in for the hydrolysis of the APP peptide bounds by γ-secretase. A sorting chaperone function like that described for SCAP, allowing transport of APP, Notch and other substrates from the ER to a more downstream compartment where γ-secretase resides, could also be envisaged. Finally, it could be that the PS are involved in a transport or maturation step of the γ-secretases themselves. The enrichment of PS in the intermediate compartment between ER and *cis*-Golgi is at least compatible with such possibilities (Annaert et al. 1999).

Conclusion

Emerging evidence firmly establishes the central role of the PS in AD and in a series of signalling pathways involved in embryonic development and the unfolded stress response. Despite tremendous efforts, it remains to be elucidated to what extent PS are directly operating as proteases, releasing protein fragments like the Notch intracellular domain that mediate transfer of information to the nucleus. Further research is anticipated to reveal the molecular basis of this intriguing activity called γ-secretase and to yield, next to the recently identified β-secretase (Hussain et al. 1999; Sinha et al. 1999; Vassar et al. 1999; Yan et al. 1999), a new clearly defined therapeutic target for AD treatment.

References

Annaert W, De Strooper B (1999) Presenilins: molecular switches between proteolysis and signal transduction. Trends Neurosci 22: 439–443

Annaert WG, Levesque L, Craessaerts K, Dierinck I, Snellings G, Westaway D, George-Hyslop PS, Cordell B, Fraser P, De Strooper B (1999) Presenilin 1 controls gamma-secretase processing of amyloid precursor protein in pre-golgi compartments of hippocampal neurons. J Cell Biol 147: 277–294

Artavanis-Tsakonas S, Rand MD, Lake RJ (1999) Notch signaling: cell fate control and signal integration in development. Science 284: 770–776

Baumeister R, Leimer U, Zweckbronner I, Jakubek C, Grunberg J, Haass C (1997) Human presenilin-1, but not familial Alzheimer's disease (FAD) mutants, facilitate Caenorhabditis elegans Notch signalling independently of proteolytic processing. Genes Funct 1: 149–159

Borchelt DR, Thinakaran G, Eckman CB, Lee MK, Davenport F, Ratovitsky T, Prada CM, Kim G, Seekins S, Yager D, Slunt HH, Wang R, Seeger M, Levey AI, Gandy SE, Copeland NG, Jenkins NA, Price DL, Younkin SG, Sisodia SS (1996) Familial Alzheimer's disease-linked presenilin 1 variants elevate Abeta1-42/1-40 ratio in vitro and in vivo. Neuron 17: 1005–1013

Brockhaus M, Grunberg J, Rohrig S, Loetscher H, Wittenburg N, Baumeister R, Jacobsen H, Haass C (1998) Caspase-mediated cleavage is not required for the activity of presenilins in amyloidogenesis and NOTCH signaling. Neuroreport 9: 1481–1486

Brown MS, Goldstein JL (1997) The SREBP pathway: regulation of cholesterol metabolism by proteolysis of a membrane-bound transcription factor. Cell 89: 331–340

Capell A, Saffrich R, Olivo JC, Meyn L, Walter J, Grunberg J, Mathews P, Nixon R, Dotti C, Haass C (1997) Cellular expression and proteolytic processing of presenilin proteins is developmentally regulated during neuronal differentiation. J Neurochem 69: 2432–2440

Capell A, Grunberg J, Pesold B, Diehlmann A, Citron M, Nixon R, Beyreuther K, Selkoe DJ, Haass C (1998) The proteolytic fragments of the Alzheimer's disease-associated presenilin-1 form heterodimers and occur as a 100–150-kDa molecular mass complex. J Biol Chem 273: 3205–3211

Chan YM, Jan YN (1999) Presenilins, processing of beta-amyloid precursor protein, and notch signaling. Neuron 23: 201–204

Citron M, Westaway D, Xia W, Carlson G, Diehl T, Levesque G, Johnson-Wood K, Lee M, Seubert P, Davis A, Kholodenko D, Motter R, Sherrington R, Perry B, Yao H, Strome R, Lieberburg I, Rommens J, Kim S, Schenk D, Fraser P, St George-Hyslop P, Selkoe DJ (1997) Mutant presenilins of Alzheimer's disease increase production of 42-residue amyloid beta-protein in both transfected cells and transgenic mice. Nature Med 3: 67–72

Conlon RA, Reaume AG, Rossant J (1995) Notch1 is required for the coordinate segmentation of somites. Development 121: 1533–1545

Culvenor JG, Maher F, Evin G, Malchiodi-Albedi F, Cappai R, Underwood JR, Davis JB, Karran EH, Roberts GW, Beyreuther K, Master C (1997) Alzheimer's disease-associated presenilin 1 in neuronal cells: evidence for localization to the endoplasmic reticulum-Golgi intermediate compartment. J Neurosci Res 49: 719–731

De Strooper B, Beullens M, Contreras B, Levesque L, Craessaerts K, Cordell B, Moechars D, Bollen M, Fraser P, George-Hyslop PS, van Leuven F (1997) Phosphorylation, subcellular localization, and membrane orientation of the Alzheimer's disease-associated presenilins. J Biol Chem 272: 3590–3598

De Strooper B, Saftig P, Craessaerts K, Vanderstichele H, Guhde G, Annaert W, Von Figura K, Van Leuven F (1998) Deficiency of presenilin-1 inhibits the normal cleavage of amyloid precursor protein. Nature 391: 387–390

De Strooper B, Annaert W, Cupers P, Saftig P, Craessaerts K, Mumm JS, Schroeter EH, Schrijvers V, Wolfe MS, Ray WJ, Goate A, Kopan R (1999) A presenilin-1-dependent gamma-secretase-like protease mediates release of Notch intracullular domain. Nature 398: 518–522

DeBose-Boyd RA, Brown MS, Li WP, Nohturfft A, Goldstein JL, Espenshade PJ (1999) Transport-dependent proteolysis of SREBP: relocation of site-1 protease from Golgi to ER obviates the need for SREBP transport to Golgi. Cell 99: 702–712

Deng G, Pike CJ, Cotman CW (1996) Alzheimer-associated presenilin-2 confers increased sensitivity to apoptosis in PC12 cells. FEBS Lett 397: 50–54

Doan A, Thinakaran G, Borchelt DR, Slunt HH, Ratovitsky T, Podlisny M, Selkoe DJ, Seeger M, Gandy SE, Price DL, Sisodia SS (1996) Protein topology of presenilin 1. Neuron 17: 1023–1030

Donoviel DB, Hadjantonakis AK, Ikeda M, Zheng H, Hyslop PS, Bernstein A (1999) Mice lacking both presenilin genes exhibit early embryonic patterning defects. Genes Dev 13: 2801–2810

Duff K, Eckman C, Zehr C, Yu X, Prada CM, Perez-tur J, Hutton M, Buee L, Harigaya Y, Yager D et al. (1996) Increased amyloid-beta42(43) in brains of mice expressing mutant presenilin 1. Nature 383: 710–713

Georgakopoulos A, Marambaud P, Efthimiopoulos S, Shioi J, Cui W, Li HC, Schutte M, Gordon R, Holstein GR, Martinelli G, Mehta P, Friedrich VL Jr, Robakis NK (1999) Presenilin-1 forms complexes with the cadherin/catenin cell-cell adhesion system and is recruited to intercellular and synapotic contacts. Mol Cell 4: 893–902

Guo Q, Sopher BL, Furukawa K, Pham DG, Robinson N, Martin GM, Mattson MP (1997) Alzheimer's presenilin mutation sensitizes neural cells to apoptosis induced by trophic factor withdrawal and amyloid beta-peptide: involvement of calcium and oxyradicals. J Neurosci 17: 4212–4222.

Haass C, De Strooper B (1999) The presenilins in Alzheimer's disease-proteolysis hold the key. Science 286: 916–919

Haass C, Selkoe DJ (1993) Cellular processing of beta-amyloid precursor protein and the genesis of amyloid beta-peptide. Cell 75: 1039–1042

Hartmann D, De Strooper BD, Saftig P (1999) Presenilin-1 deficiency leads to loss of Cajal-Retzius neurons and cortical dysplasia similar to human type 2 lissencephaly. Curr Biol 9: 719–727

Herreman A, Hartmann D, Annaert W, Saftig P, Craessaerts K, Serneels L, Umans L, Schrijvers V, Checler F, Vanderstichele H, Baekelandt V, Dressel R, Cupers P, Huylebroeck D, Zwijsen A, Van Leuven F, De Strooper B (1999) Presenilin 2 deficiency causes a mild pulmonary phenotype and no changes in amyloid precursor protein processing but enhances the embryonic lethal phenotype of presenilin 1 deficiency. Proc Natl Acad Sci USA 96: 11872–11877

Hussain I, Powell D, Howlett DR, Tew DG, Meek TD, Chapman C, Gloger IS, Murphy KE, Southan CD, Ryan DM, Smiths TS, Simmons DL, Walsh FS, Dingwall C, Christie G (1999) Identification of a novel aspartic protease (Asp 2) as beta-secretase. Mol Cell Neurosci 14: 419–427

Janicki S, Monteiro MJ (1997) Increased apoptosis arising from increased expression of the Alzheimer's disease-associated presenilin-2 mutation (N141I). J Cell Biol 139: 485–495

Katayama T, Imaizumi K, Sato N, Miyoshi K, Kudo T, Hitomi J, Morihara T, Yoneda T, Gomi F, Mori Y, Nakano Y, Takeda J, Tsuda T, Itoyama Y, Murayama O, Takashima A, St George-Hyslop P, Takeda M, Tohyama M, Imaizumi K (1999) Presenilin-1 mutations downregulate the signalling pathway of the unfolded-protein response. Nat use Cell Biol 1: 479–485

Kovacs DM, Fausett HJ, Page KJ, Kim TW, Moir RD, Merriam DE, Hollister RD, Hallmark OG, Mancini R, Felsenstein KM, Hyman BT, Tanzi RE, Wasco W(1996) Alzheimer-associated presenilins 1 and 2: neuronal expression in brain and localization to intracellular membranes in mammalian cells. Nature Med 2: 224–229

Lah JJ, Heilman CJ, Nash NR, Rees HD, Yi H, Counts SE, Levey AI (1997) Light and electron microscopic localization of presenilin-1 in primate brain. J Neurosci 17: 1971–1980

Lehmann S, Chiesa R, Harris DA (1997) Evidence for a six-transmembrane domain structure of presenilin 1. J Biol Chem 272: 12047–12051

Leimer U, Lun K, Romig H, Walter J, Grunberg J, Brand M, Haass C (1999) Zebrafish (*Danio rerio*) presenilin promotes aberrant amyloid beta-peptide production and requires a critical aspartate residue for its function in amyloidogenesis. Biochemistry 38: 13602–13609

Lemere CA, Lopera F, Kosik KS, Lendon CL, Ossa J, Saido TC, Yamaguchi H, Ruiz A, Martinez A, Madrigal L, Hincapie L, Arango JC, Anthony DC, Koo EH, Goate AM, Selkoe DJ, Arango JC (1996) The E280A presenilin 1 Alzheimer mutation produces increased A beta 42 deposition and severe cerebellar pathology. Nature Med 2: 1146–1150

Levesque G, Yu G, Nishimura M, Zhang DM, Levesque L, Yu H, Xu D, Liang Y, Rogaeva E, Ikeda M, Duthie M, Murgolo N, Wang L, VanderVere P, Bayne ML, Strader CD, Rommens JM, Fraser PE, St George-Hyslop P (1999) Presenilins interact with armadillo proteins including neural-specific plakophilin-related protein and beta-catenin. J Neurochem 72: 999–1008

Levitan D, Greenwald I (1995). Facilitation of lin-12-mediated signalling by sel-12, a *Caenorhabditis elegans* S182 Alzheimer's disease gene. Nature 377: 351–354

Levy-Lahad E, Wasco W, Poorkaj P, Romano DM, Oshima J, Pettingell WH, Yu CE, Jondro PD, Schmidt SD, Wang K, Crowley AC, Fu YH, Guenette SY, Galas D, Nemens E, Wijsman EM, Bird TD, Schellenberg GD, Tanzi RE (1995) Candidate gene for the chromosome 1 familial Alzheimer's disease locus. Science 269: 973–977

Li X, Greenwald I (1996) Membrane topology of the *C. elegans* SEL-12 presenilin. Neuron 17: 1015–1021

Li X, Greenwald I (1998) Additional evidence for an eight-transmembrane-domain topology for *Caenorhabditis elegans* and human presenilins. Proc Natl Acad Sci USA 95: 7109–7114

Murayama M, Tanaka S, Palacino J, Murayama O, Honda T, Sun X, Yasutake K, Nihonmatsu N, Wolozin B, Takashima A (1998) Direct association of presenilin-1 with beta-catenin. FEBS Lett 433: 73–77

Murphy MP, Hickman LJ, Eckman CB, Uljon SN, Wang R, Golde TE (1999) gamma-Secretase, evidence for multiple proteolytic activities and influence of membrane positioning of substrate on generation of amyloid beta peptides of varying length. J Biol Chem 274: 11914–11923

Naruse S, Thinakaran G, Luo JJ, Kusiak JW, Tomita T, Iwatsubo T, Qian X, Ginty DD, Price DL, Borchelt DR et al. (1998) Effects of PS1 deficiency on membrane protein trafficking in neurons. Neuron 21: 1213–1221

Niwa M, Sidrauski C, Kaufman RJ, Walter P (1999) A role for presenilin-1 in nuclear accumulation of Ire1 fragments and induction of the mammalian unfolded protein response. Cell 99: 691–702

Nohturfft A, DeBose-Boyd RA, Scheek S, Goldstein JL, Brown MS (1999) Sterols regulate cycling of SREBP cleavage-activating protein (SCAP) between endoplasmic reticulum and Golgi. Proc Natl Acad Sci USA 96: 11235–11240

Rawson RB, Zelenski NG, Nijhawan D, Ye J, Sakai J, Hasan MT, Chang TY, Brown MS, Goldstein JL (1997) Complementation cloning of S2P, a gene encoding a putative metalloprotease required for intramembrane cleavage of SREPBs. Mol Cell 1: 47–57

Ray WJ, Yao M, Mumm J, Schroeter EH, Saftig P, Wolfe M, Selkoe DJ, Kopan R, Goate AM (1999) Cell surface presenilin-1 participates in the gamma-secretase-like proteolysis of notch. J Biol Chem 274: 36801–36807

Rogaev EI, Sherrington R, Rogaeva EA, Levesque G, Ikeda M, Liang Y, Chi H, Lin C, Holman K, Tsuda T, Mar L, Sorbi S, Nacmias B, Piacentini S, Amaducci L, Chumakov I, Cohen D, Lannfelt L, Fraser PE, Rommens JM, St George-Hyslop P (1995) Familial Alzheimer's disease in kindreds with missense mutations in a gene on chromosome 1 related to the Alzheimer's disease type 3 gene. Nature 376: 775–778

Ross SL, Martin F, Somonet L, Jacobsen F, Deshpande R, Vassar R, Bennett B, Luo Y, Wooden S, Hu S, Citron M, Burgess TL (1998) Amyloid precursor protein processing in sterol regulatory element-binding protein site 2 protease-deficient Chinese hamster ovary cells. J Biol Chem 273: 15309–15312

Sakai J, Rawson RB, Espenshade PJ, Cheng D, Seegmiller AC, Goldstein JL, Brown MS (1998) Molecular identification of the sterol-regulated luminal protease that cleaves SREBPs and controls lipid composition of animal cells. Mol Cell 2: 505–514

Scheuner D, Eckman C, Jensen M, Song X, Citron M, Suzuki N, Bird TD, Hardy J, Hutton M, Kukull W et al. (1996) Secreted amyloid beta-protein similar to that in the senile plaques of Alzheimer's disease is increased in vivo by the presenilin 1 and 2 and APP mutations linked to familial Alzheimer's disease. Nature Med 2: 864–870

Schroeter EH, Kisslinger JA, Kopan R (1998) Notch-1 signalling requires ligand-induced proteolytic release of intracellular domain. Nature 393: 382–386

Shen J, Bronson RT, Chen DF, Xia W, Selkoe DJ, Tonegawa S (1997) Skeletal and CNS defects in Presenilin-1-eficient mice. Cell 89: 629–639

Sherrington R, Rogaev EI, Liang Y, Rogaeva EA, Levesque G, Ikeda M, Chi H, Lin C, Li G, Holman K, Tsuda T, Mar L, Foncin JF, Bruni AC, Montesi MP, Sorbi S, Rainero I, Pinessi L, Nee L, Chumakov I, Pollen D, Brookes A, Sanseau P, Polinsky RJ, Wasco W, Da Silva HAR, Haines JL, Pericak-Vance MA, Tanzi RE, Roses AD, Fraser PE, Rommens JM, St George-Hyslop P (1995) Clining of a gene bearing missense mutations in early-onset familial Alzheimer's disease. Nature 375: 754–760

Sinha S, Anderson JP, Barbour R, Basi GS, Caccavello R, Davis D, Doan M, Dovey HF, Frigon N, Hong H, Jacobson-Croak K, Jewett N, Keim P, Knops J, Lieberburg I, Power M, Tan H, Tatsuno G, Tung J, Schenk D, Seubert P, Suomensaari SM, Wang S, Walker D, Zhao J, McConlogue L, John V (1999) Purification and cloning of amyloid precursor protein beta-secretase from human brain. Nature 402: 537–540

Song W, Nadeau P, Yuan M, Yang X, Shen J, Yankner BA (1999) Proteolytic release and nuclear translocation of Notch-1 are induced by presenilin-1 and impaired by pathogenic presenilin-1 mutations. Proc Natl Acad Sci USA 96: 6959–6963

Steiner H, Duff K, Capell A, Romig H, Grim MG, Lincoln S, Hardy J, Yu X, Picciano M, Fechteler K, Citron M, Kopan R, Pesold B, Keck S, Baader M, Tomita T, Iwatsubo T, Baumeister R, Haass C (1999) A loss of function mutation of presenilin-2 interferes with amyloid beta-peptide production and notch signaling. J Biol Chem 274: 28669–28673

Struhl G, Greenwald I (1999) Presenilin is required for activity and nuclear access of Notch in Drosophila. Nature 398: 522–525

Swiatek PJ, Lindsell CE, del Amo FF, Weinmaster G, Gridley T (1994) Notch 1 is essential for postimplantation development in mice. Genes Dev 8: 707–719

Tesco G, Kim TW, Diehlmann A, Beyreuther K, Tanzi RE (1998) Abrogation of the presenilin 1/beta-catenin interaction and preservation of the heterodimeric presenilin 1 complex following caspase activation. J Biol Chem 273: 33909–33914

Thinakaran G, Borchelt DR, Lee MK, Slunt HH, Spitzer L, Kim G, Ratovitsky T, Davenport F, Nordstedt C, Seeger M, Hardy J, Levey AI, Gandy SE, Jenkins NA, Copeland NG, Price DL, Sisodia SS (1996) Endoproteolysis of presenilin 1 and accumulation of processed derivatives in vivo. Neuron 17: 181–190

Tomita T, Maruyama K, Saido TC, Kume H, Shinozaki K, Tokuhiro S, Capell A, Walter J, Grunberg J, Haass C, Iwatsubo T, Obaka K (1997) The presenilin 2 mutation (N141I) linked to familial Alzheimer disease (Volga German families) increases the secretion of amyloid beta protein ending at the 42nd (or 43rd) residue. Proc Natl Acad Sci USA 94: 2025–2030

Vassar R, Bennett BD, Babu-Khan S, Kahn S, Mendiaz EA, Denis P, Teplow DB, Ross S, Amarante P, Loeloff R et al. (1999) Beta-secretase cleavage of Alzheimer's amyloid precursor protein by the trans-membrane aspartic protease BACE. Science 286: 735–741

Vito P, Ghayur T, Adamio LD (1997) Generation of anti-apoptotic presenilin-2 polypeptides by alternative transcription, proteolysis, and caspase-3 cleavage. J Biol Chem 272: 28315–28320

Walter J, Capell A, Grunberg J, Pesold B, Schindzielorz A, Prior R, Podlisny MB, Fraser P, Hyslop PS, Sekoe DJ et al. (1996) The Alzheimer's disease-associated presenilins are differentially phosphory-lated proteins located predominantly within the endoplasmic reticulum. Mol Med 2: 673–691

Wolfe MS, De Los Angeles J, Miller DD, Xia W, Selkoe DJ (1999a) Are presenilins intramembrane-cleaving proteases? Implications for the molecular mechanism of Alzheimer's disease. Biochemis-try 38: 11223–11230

Wolfe MS, Xia W, Moore CL, Leatherwood DD, Ostaszewski B, Rahgmati T, Donkor IO, Selkoe DJ (1999b) Peptidomimetic probes and molecular modeling suggest that Alzheimer's gamma-secretase is an intramembrane-cleaving aspartyl protease. Biochemistry 38: 4720–4727

Wolfe MS, Xia W, Ostaszewski BL, Diehl TS, Kimberly WT, Selkoe DJ (1999c) Two transmembrane aspartates in presenilin-1 required for presenelin endoproteolysis and gamma-secretase activity. Nature 398: 513–517

Wolozin B, Iwasaki K, Vito P, Ganjei JK, Lacana E, Sunderland T, Zhao B, Kusiak JW, Wasco W, Adamio LD (1996) Participation of presenilin 2 in apoptosis: enhanced basal activity conferred by an Alzheimer mutation. Science 273: 1710–1713

Wong PC, Zheng H, Chen H, Becher MW, Sirinathsinghji DJ, Trumbauer ME, Chen HY, Price DL, Van der Ploeg LH, Sisodia SS (1997) Presenilin 1 is required for Notch 1 and DII1 expression in the par-axial mesoderm. Nature 387: 288–292

Xia W, Zhang J, Kholodenko D, Citron M, Podlisny MB, Teplow DB, Haass C, Seubert P, Koo EH, Sel-koe DJ (1997). Enhanced production and oligomerization of the 42-residue amyloid beta-protein by Chinese hamster ovary cells stably expressing mutant presenilins. J Biol Chem 272: 7977–7982

Yan R, Bienkowski MJ, Shuck ME, Miao H, Tory MC, Pauley AM, Brashier JR, Stratman NC, Mathews WR, Buhl AE et al. (1999) Membrane-anchored aspartyl protease with Alzheimer's disease beta-secretase activity. Nature 402: 533–537

Ye Y, Lukinova N, Fortini ME (1999) Neurogenic phenotypes and altered Notch processing in Droso-phila Presenilin mutants. Nature 398: 525–529

Yu G, Chen F, Levesque G, Nishimura M, Zhang DM, Levesque L, Rogaeva E, Xu D, Liang Y, Duthie M et al. (1998) The presenilin 1 protein is a component of a high molecular weight intracellular com-plex that contains beta-catenin. J Biol Chem 273: 16470–16475

Zhang Z, Hartmann H, Do VM, Abramowski D, Sturchler-Pierrat C, Staufenbiel M, Sommer B, van de Wetering M, Vlevers H, Saftig P et al. (1998) Destabilization of beta-catenin by mutations in presenilin-1 potentiates neuronal apoptosis. Nature 395: 698–702.

Zhou J, Liyanage U, Medina M, Ho C, Simmons AD, Lovett M, Kosik KS (1997) Presenilin 1 interaction in the brain with a novel member of the Armadillo family. Neuroreport 8: 1489–1494

ApoE Receptors in the Brain:
Novel Signaling Pathways with Potential Relevance for Alzheimer's Disease

J. Herz, U. Beffert, T. Hiesberger, and M. Gotthardt

Summary

Apolipoprotein E (ApoE) is an intrinsic component of lipoproteins and is known to transport cholesterol and other lipids through the circulation and between cells. ApoE binds to a family of cell surface receptors, the LDL receptor gene family. This class of endocytic receptors mediates the internalization of the lipoprotein particle, followed by transport of the lipids to the lysosomes. Several years ago, one isoform of ApoE, ApoE4, was reported to be genetically associated with late-onset Alzheimer's disease. The biochemical mechanism by which ApoE4 predisposes its carriers to this debilitating neurodegenerative condition remains largely an enigma. We have investigated whether ApoE receptors on neurons are involved in the molecular pathogenic process. LRP, a member of the LDL receptor gene family, binds the amyloid precursor protein (APP) on its extracellular domain and is also connected to the cytoplasmic tail of APP by the adaptor protein FE65. This two-point interaction suggests that LRP may affect the subcellular localization, routing, and possibly processing of APP. Recently, we also described a novel signaling pathway by which two other members of the family, the VLDL receptor and the ApoE receptor 2, serve to transmit a positional cue across the membrane to migrating neurons. The signaling molecule, Reelin, binds to the receptor ectodomains and induces tyrosine phosphorylation of the intracellular adaptor protein Disabled-1. Disruption of this ApoE receptor-mediated signal transduction pathway results in hyperphosphorylation of the microtubule stabilizing protein tau, a prerequisite for the formation of neurofibrillary tangles in humans afflicted with Alzheimer's disease. These findings tie neuronal ApoE receptors directly to two of the prominent pathological changes that are diagnostic for the disease: amyloid plaques, caused by abnormal processing of APP, and neurofibrillary tangles, caused by hyperphosphorylation of tau. We conclude that ApoE receptors regulate essential cellular signaling pathways in neurons and that defects in these pathways may accelerate the onset of Alzheimer's disease.

Evolutionary Conservation of the LDL Receptor Gene Family

Five members of the low density (LDL) receptor gene family are presently known in mammalian organisms. They include the LDL receptor (Brown and Goldstein 1986), the LDL receptor-related protein (LRP; Herz et al. 1988), megalin (Saito

Research and Perspectives in Alzheimer's Diseases
Beyreuther/Christen/Masters (Eds.)
Neurodegenerative Disorders
© Springer-Verlag Berlin Heidelberg 2001

et al. 1994), the very low density lipoprotein (VLDL) receptor (Takahashi et al. 1992), and the apolipoprotein E (ApoE) receptor-2 (ApoER2; Kim et al. 1996; Novak et al. 1996). Both on the level of gene as well as protein structure the members of the family are closely related to each other. The extracellular domains are made up of three basic structural modules: cysteine-rich, negatively charged ligand-binding type repeats that are also present in the terminal complement components, epidermal growth factor-like repeats, and YWTD repeats (Krieger and Herz 1994). All receptors are anchored in the plasma membrane by a single membrane spanning segment that is followed by a short cytoplasmic tail between approximately 50 and 200 amino acids in length. A conserved tetra amino acid motif, the so-called NPxY motif, is present in one (LDL receptor, VLDL receptor and ApoER2), two (LRP), or three (megalin) copies in the tails. This motif has been shown to mediate the interaction of the LDL receptor with the coated pit endocytosis machinery. However, as we will see, it has other important functions in signal transduction events in which the LDL receptor family is involved.

The gene family arose very early during evolution. A member of the family is present in the primitive multicellular organism *Caenorhabditis elegans* (Yochem and Greenwald 1993; Yochem et al. 1999). The remarkable structural conservation of this gene – it is virtually identical in the arrangement of its structural elements to its modern counterpart, megalin – and the absence of the gene family from the genome of the monocellular eukaryote *Saccharomyces cervisiae*, strongly suggest that this gene family originally evolved to fulfill critical functions in the organization of multicellular organisms (Willnow et al. 1999).

Functions in Lipoprotein Metabolism

The LDL receptor gene family received its name from the founding member of the family. The LDL receptor was originally identified because it is mutated in human patients carrying the disease "Familial Hypercholesterolemia" (FH). Its physiological functions seem to be limited to the removal of LDL, which is the major carrier of cholesterol in the plasma, from the circulation. The receptor achieves this by the process of receptor-mediated endocytosis. Binding of ligand, in this case LDL, to the extracellular domain of the receptor is followed by periodic clustering of the ligand/receptor complex in coated pits at the cell surface. These pits pinch off to form coated vesicles that eventually unload the bound ligand in the lysosomes where it is digested. LDL and other lipoproteins are essentially liposomes in which protein components, the so-called apoproteins, are embedded. It is these apoproteins that mediate the binding of the lipoprotein particle to cell surface receptors such as the LDL receptor. The LDL particle harbors a single apoprotein called ApoB100, which binds to the ligand binding domain of the receptor. A defect in the LDL receptor results in a dramatically increased total plasma cholesterol level and predisposition to coronary artery disease. Homozygous FH patients typically develop subcutaneous xanthomas even before puberty and usually die from coronary artery disease before the age of 20.

Because of the high degree of structural conservation between the members of the LDL receptor gene family, it was originally surmised that all these genes also coded for lipoprotein receptors that are involved in various aspects of lipoprotein transport between tissues and cells. However, so far a physiologically important function in lipoprotein transport has only been demonstrated for one other member of the family, LRP. This gargantuan receptor, which weighs in at around 600 kDa, functions in concert with LDL receptor in the removal of the carriers of dietary lipids, the chylomicron remnants, from the circulation (Rohlmann et al. 1998). In this role, both receptors bind the remnant lipoproteins via an interaction with ApoE that is present on the surface of the remnants in multiple copies. Although all members of the family can bind ApoE on their extracellular domains, genetic defects in these receptors do not appear to affect systemic lipoprotein transport in any readily detectable way. Indeed, functions unrelated to lipoprotein metabolism have recently been identified for these receptors that begin to explain their high degree of evolutionary conservation and their quite distinct tissue distribution pattern (Willnow et al. 1999).

Functions in the Regulation of Cell Surface Protease Activity

The first indication that the physiological functions of LDL receptor family members are not restricted to lipoprotein metabolism arose from the findings by groups led by Strickland (Strickland et al. 1990) and Gliemann (Kristensen et al. 1990), who showed that LRP was identical to the well-studied receptor for α_2-macroglublin (α_2M). α_2M is an abundant plasma protein that binds and traps a broad range of proteases. Interaction with these proteases leads to cleavage of so-called bait region in α_2M, which in turn results in a conformational change of the protein that exposes a high affinity binding site for its receptor. Removal of the α_2M-protease complexes from the liver is rapid, resulting in hepatic uptake and lysosomal degradation. Following this discovery, it was rapidly shown that LRP also interacts with other protease inhibitors, such as plasminogen activator inhibitor, but also with the active proteases tissue-type and urokinase-type plasminogen activator (Bu et al. 1992; Nykjaer et al. 1992; Orth et al. 1992). A role for this receptor is firmly established in the homeostasis of several plasma proteases that are mostly involved in coagulation and fibrinolysis. As will be briefly discussed below, interaction of receptors with proteases and protease inhibitor complexes can result in the activation of cellular signaling events (Webb et al. 1999, 2000).

Functions in Vitamin Metabolism

Another physiological function for the LDL receptor that has recently emerged is its role in the homeostasis of vitamins. Although a role in the transport of riboflavin into the growing egg in birds had long been suspected to involve a member of the LDL receptor family (Mac Lachlan et al. 1994), the physiological importance of such transport processes in mammalian organisms had not been realized until the detailed analysis of knockout mice lacking the family member megalin. Megalin is highly expressed on the apical surface of many villous resorptive epithelia, such as the gut, the neuroepithelium, the yolk sac, and the cells of the proximal tubulus in the kidney. Megalin-deficient mice show multiple developmental defects. The most prominent defect affects the formation of the brain, causing a malformation commonly referred to as holoprosencephaly (Willnow et al. 1996). Because of this developmental defect, most megalin-deficient mice die shortly after birth. However, a small percentage of these mice survive. This has made it possible to investigate the role of megalin in the kidney. Because this protein is expressed at very high levels on the surface of the tubular cells which is facing the lumen of the tubulus, it had been hypothesized that megalin might be involved in the reabsorption of proteins that are present in the primary glomerular filtrate. Analysis of proteins that are preferentially lost in the urine of megalin knockout mice revealed that two of these proteins are carriers for lipophilic vitamins, retinol binding protein (a carrier for vitamin A) and vitamin D binding protein. As a consequence of the loss of these vitamin carriers through the urine, megalin-deficient mice are severely deficient in vitamin D and show massive defects in bone mineralisation (Nykjaer et al. 1999). Not much is known at present about possible functions of the LDL receptor gene family in the transport of vitamins to peripheral tissues in the body, although some evidence suggests that high affinity receptors for the uptake of vitamin carrier complexes are present in tissues with a high demand for steroid hormones, such as the gonads and also some types of cancer (Willnow et al. 1999).

Functions in Cellular Signal Transduction

The rapid discovery of numerous, biologically diverse ligands that interact with the extracellular domains of LDL receptor family members has painted a bewildering picture of the possible physiological functions in which this gene family may be involved. Furthermore, the broad range of phenotypes found in humans and animals lacking one or more of the receptors has been difficult to explain on the basis of functions that are limited to the endocytic uptake of ligands bound to the extracellular domains of the receptors and their lysosomal delivery. Based on these observations we hypothesized that the LDL receptor gene family may have physiological roles that transcend its well-established role in mediating endocytosis. As many developmental defects are caused by an inability of cells to relay to or receive proper signals from their surroundings, we considered the possibility that the receptors might be involved in cellular signal transduction events. In most cases of cell surface signaling receptors, this process involves the cytoplasmic tail. We thus searched for cytoplasmic proteins that might bind to the receptor tails and would be capable of relaying a signal to established intracellular signaling pathways. Our initial screens revealed two proteins Dab 1 and FE65, that bound tightly to the tails of the LDL receptor and the LRP (Trommsdorff et al. 1998). Dab 1 and FE65 both contain protein interaction domains of a type known to bind to ‚NPxY' motifs like those present in the tails of the LDL receptor gene family. Dab 1 was already known to function in a signaling pathway by which migrating neurons receive an important positional cue (Howell et al. 1997, 1999). This pathway involves the large modular protein Reelin, which is secreted by a specialized class of neurons, the Cajal-Retzius neurons, on the surface of the developing brain and regulates the lamination of the cortex (Rice and Curran 1999). The receptors that relay the Reelin signal across the plasma membrane of the migrating neurons had not been identified, however. Analysis of knockout mice lacking two members of the LDL receptor gene family, the VLDL receptor and the ApoER2, revealed a phenotype that was indistinguishable from that of mice lacking Reelin or Dab 1 (Trommsdorff et al. 1999). This finding suggested that these two receptors function in a partially overlapping fashion in the transmission of this critical signal during embryonic brain development. Subsequent studies then rapidly showed that Reelin does indeed bind to the extracellular domains of VLDL receptor and ApoER2, but not to the LDL receptor (D'Arcangelo et al. 1999; Hiesberger et al. 1999), and that this binding is necessary for tyrosine phosphorylation of Dab 1. These studies for the first time established a firm genetic and biochemical basis for functions of LDL receptor family members in cellular signal transduction.

Recently, Gonias and colleagues have independently established a signaling pathway that involves the interaction of urokinase type plasminogen activator with a specific GPI linked cell surface receptor and integrins, but apparently also involving members of the LDL receptor gene family (Webb et al. 2000, 1999).

Implications for Alzheimer's Disease

How do the seemingly unrelated functions of the LDL receptor gene family in such diverse processes as lipoprotein metabolism, protease regulation, vitamin metabolism and neuronal migration suggest a role for this ancient gene family in the development of a devastating neurodegenerative disorder? Genetic association studies have connected several genes that functionally interact with the LDL receptor gene family with the pathologic process that leads to Alzheimer's disease. The strongest association has been observed for one of the ligands of the LDL receptor gene family, ApoE. In particular one isoform of ApoE, ApoE4, strongly predisposes its carriers to late-onset Alzheimer's disease (Schmechel et al. 1993). A probable assumption is that ApoE mediates this effect through its receptors, which are abundantly expressed on the surface of all neurons in the brain. Furthermore, several years ago the amyloid precursor protein (APP), an integral membrane protein expressed by glial cells and neurons in the brain, was shown to interact directly with the LRP via its extracellular domain (Kounnas et al. 1995). Abnormal proteolytic processing of APP gives rise to the abnormal amyloid deposits in patients afflicted with Alzheimer's disease. In addition to this extracellular interaction, APP and LRP appear to also be connected on the cytoplasmic side by the scaffolding protein FE65. These findings suggest that the trafficking and thus the processing of APP might be regulated by its interaction with members of the LDL receptor gene family, a process that could in principle be modulated by ApoE.

Another hallmark of Alzheimer's disease are the neurofibrillary tangles that form inside afflicted neurons. These tangles are caused by the hyperphosphorylated form of the microtubule stabilizing protein tau. Abnormal phosphorylation of tau causes it to dissociate from the microtubules, which in turn destabilizes the microtubular network affecting axonal transport and thus neuronal cell function in general. The signals that cause the normally delicately balanced system of kinases and phosphatases that regulate the phosphorylation state of the tau protein to go awry are only now beginning to emerge. Two of the genes involved in tau phosphorylation are the kinase cdk5 and its regulator p35 (Chae et al. 1997; Ohsihima et al. 1996; Patrick et al. 1999). Mice lacking either of these genes exhibit neuronal migration defects that resemble those seen in animals carrying a defect in the Reelin signaling pathway, including mutations in the two neuronal ApoE receptors, VLDL receptor and ApoER2. We have therefore tested the hypothesis whether mutations in this signaling pathway impact on the phosphorylation state of the tau protein. Indeed, mice in which this signaling pathway is dysfunctional, either because they lack Reelin or because they carry mutations in VLDL receptor and ApoER2, have greatly elevated levels of abnormally phosphorylated tau protein in their brains (Hiesberger et al. 1999). Thus, two of the hallmarks of Alzheimer's disease, the amyloid lesions on one hand and neurofibrillary tangles on the other, can be linked by direct physical, functional or genetic interaction to members of the LDL receptor gene family in the brain. These findings now form a basis for a detailed biochemical and genetic analysis of the molecular processes by which neurons degenerate in Alzheimer's disease. They also suggest that maintenance of neuronal signal input via ApoE receptors is vital to the well-being of the mind.

References

Brown MS, Goldstein JL (1986) A receptor-mediated pathway for cholesterol homeostasis. Science 232: 34–47

Bu G, Williams S. Strickland DK, Schwartz AL (1992) Low density lipoprotein receptor-related protein/ α_2-macroglobulin receptor is an hepatic receptor for tissue-type plasminogen activator. Proc Natl Acad Sci USA 89: 7427–7431

Chae T, Kwon YT, Bronson R, Dikkes P, Li E, Tsai LH (1997) Mice lacking p35, a neuronal specific activator of Cdk5, display cortical lamination defects, seizures, and adult lethality. Neuron 18: 29–42

D'Arcangelo G, Homayouni R, Keshvara L, Rice DS, Sheldon M, Curran T (1999) Reelin is a ligand for lipoprotein receptors. Neuron 24: 471–479

Herz J, Hamann U, Rogne S, Myklebost O, Gausepohl H, Stanley KK (1988) Surface location and high affinity for calcium of a 500-kd liver membrane protein closely related to the LDL-receptor suggest a physiological role as lipoprotein receptor. Embo J 7: 4119–4127

Hiesberger T, Trommsdorff M, Howell BW, Goffinet A, Mumby MC, Cooper JA, Herz J (1999) Direct binding of Reelin to VLDL receptor and ApoE receptor 2 induces tyrosine phosphorylation of disabled-1 and modulates tau phosphorylation. Neuron 24: 481–489

Howell BW, Hawkes R, Soriano P, Cooper JA (1997) Neuronal position in the developing brain is regulated by mouse disabled-1. Nature 389: 733–737

Howell BW, Herricks TM, Cooper JA (1999) Reelin-induced tyrosine phosphorylation of disabled-1 during neuronal positioning. Genes Dev 13: 643–648

Kim DH, Iijima H, Goto K, Sakai J, Ishii H, Kim HJ, Suzuki H, Kondo H, Saeki S, Yamamoto T (1996) Human apolipoprotein E receptor 2. A novel lipoprotein receptor of the low density lipoprotein receptor family predominantly expressed in the brain. J Biol Chem 271: 8373–8380

Kounnas MZ, Moir RD, Rebeck GW, Bush AI, Argraves WS, Tanzi RE, Hyman BT, Strickland DK (1995) LDL receptor-related protein, a multifunctional ApoE receptor, binds secreted β-amyloid precursor protein and mediates its degradation. Cell 82: 331–340

Krieger M, Herz J (1994) Structures and functions of multiligand lipoprotein receptors: macrophage scavenger receptors and LDL receptor-related protein (LRP). Ann Rev Biochem 63: 601–637

Kristensen T, Moestrup SK, Gliemann J, Bendtsen L, Sand O, Sottrup-Jensen L (1990) Evidence that the newly cloned low-density-lipoprotein receptor related protein (LRP) is the α2-macroglobulin receptor. FEBS Lett 276: 151–155

Mac Lachlan I, Nimpf J, Schneider WJ (1994) Avian riboflavin binding protein binds to lipoprotein receptors in association with vitellogenin. J Biol Chem 269: 24127–24132

Novak S, Hiesberger T, Schneider WJ, Nimpf J (1996) A new low density lipoprotein receptor homologue with 8 ligand binding repeats in brain of chicken and mouse (published erratum appears in J Biol Chem 1996 Oct 25, 271(43): 27188). J Biol Chem 271: 11732–11736

Nykjaer A, Petersen CM, Moller B, Jensen PA, Moestrup SK, Holtet TL, Etzerodt M, Thogersen HC, Munch M, Andreasen PA, Gliemann J (1992) Purified α_2-macroglobulin receptor/LDL receptor-related protein binds urokinase:plasminogen activator inhibitor type-1 complex. J Biol Chem 267: 14543–14546

Nykjaer A, Dragun D, Walther D, Vorum H, Jacobsen C, Herz J, Melsen F, Christensen EI, Willnow TE (1999) An endocytic pathway essential for renal uptake and activation of the steroid 25-(OH) vitamin D3 (in process citation). Cell 96: 507–515

Ohshima T, Ward JM, Huh CG, Longenecker G, Veeranna Pant HC, Brady RO, Martin LJ, Kulkarni AB (1996) Targeted disruption of the cyclin-dependent kinase 5 gene results in abnormal corticogenesis, neuronal pathology and perinatal death. Proc Natl Acad Sci USA 93: 11173–11178

Orth K, Madison EL, Gething MJ, Sambrook JF, Herz J (1992) Complexes of tissue-type plasminogen activator and its serpin inhibitor plasminogen-activator inhibitor type 1 are internalized by means of the low density lipoprotein receptor-related protein/alpha 2-macroglobulin receptor. Proc Natl Acad Sci USA 89: 7422–7426

Patrick GN, Zukerberg L, Nikolic M, de la Monte S, Dikkes P, Tsai LH (1999) Conversion of p35 to p25 deregulates Cdk5 activity and promotes neurodegeneration. Nature 402: 615–622

Rice DS, Curran T (1999) Mutant mice with scrambled brains: understanding the signaling pathways that control cell positioning in the CNS. Genes Dev 13: 2758–2773

Rohlmann A, Gotthardt M, Hammer RE, Herz J (1998) Inducible inactivation of hepatic LRP gene by cre-mediated recombination confirms role of LRP in clearance of chylomicron remnants. J Clin Invest 101: 689–695

Saito A, Pietromonaco S, Loo AK, Farquhar MG (1994) Complete cloning and sequencing of rat gp330/"megalin," a distinctive member of the low density lipoprotein receptor gene family. Proc Natl Acad Sci USA 91: 9725–9729

Schmechel DE, Saunders AM, Strittmatter WJ, Crain BJ, Hulette CM, Joo SH, Pericak-Vance MA, Goldgaber D, Roses AD (1993) Increased amyloid beta-peptide deposition in cerebral cortex as a consequence of apolipoprotein E genotype in late-onset Alzheimer's disease. Proc Natl Acad Sci USA 90: 9649–9653

Strickland DK, Ashcom JD, Williams S, Burgess WH, Migliorini M, Argraves WS (1990) Sequence identity between the α2-macroglobulin receptor and low density lipoprotein receptor-related protein suggests that this molecule is a multifunctional receptor. J Biol Chem 265: 17401–17404

Takahashi S, Kawarabayasi Y, Nakai T, Sakai J, Yamamoto T (1992) Rabbit very low density lipoprotein receptor: a low density lipoprotein receptor-like protein with distinct ligand specificity. Proc Natl Acad Sci USA 89: 9252–9256

Trommsdorff M, Borg JP, Margolis B, Herz J (1998) Interaction of cytosolic adaptor proteins with neuronal apolipoprotein E receptors and the amyloid precursor protein. J Biol Chem 273: 33556–33560

Trommsdorff M, Gotthardt M, Hiesberger T, Shelton J, Stockinger W, Nimpf J, Hammer RE, Richardson JA, Herz J (1999) Reeler/Disabled-like disruption of neuronal migration in knockout mice lacking the VLDL receptor and ApoE receptor 2. Cell 97: 689–701

Webb DJ, Nguyen DH, Sankovic M, Gonias SL (1999) The very low density lipoprotein receptor regulates urokinase receptor catabolism and breast cancer cell motility in vitro. J Biol Chem 274: 7412–7420

Webb DJ, Nguyen DH, Gonias SL (2000) Extracellular signal-regulated kinase functions in the urokinase receptor-dependent pathway by which neutralization of low density lipoprotein receptor-related protein promotes fibrosarcoma cell migration and matrigel invasion (in process citation). J Cell Sci 113: 123–134

Willnow TE, Hilpert J, Armstrong SA, Rohlmann A, Hammer RE, Burns DK, Herz J (1996) Defective forebrain development in mice lacking gp330/megalin. Proc Natl Acad Sci USA 93: 8460–8464

Willnow TE, Nykjaer A, Herz J (1999) Lipoprotein receptors: new roles for ancient proteins. Nature Cell Biol 1: E157–E162

Yochem J, Greenwald I (1993) A gene for a low density lipoprotein receptor-related protein in the nematode Caenorhabditis elegans. J Biol Chem 268: 13002–13009

Yochem J, Tuck S, Greenwald I, Han M (1999) A gp330/megalin-related protein is required in the major epidermis of Caenorhabditis elegans for completion of molting. Development 126: 597–606

Homeoprotein Intercellular Tansport: Mechanisms, Significance and Applications

B. Allinquant, G. Mainguy, and A. Prochiantz

Summary

In the past few years we have demonstrated that homeoprotein transcription factors can be secreted and internalized by live cells. The third helix of the homeodomain (DNA-binding domain) is necessary and sufficient for internalization, whereas secretion requires a different domain with nuclear export sequence properties. Homeoprotein secretion requires nuclear export followed by an association with cholesterol-rich rafts, and homeoprotein internalization correlates with the formation of inverted micelles and is independent of classical endocytosis. In addition to the possible important physiological functions of this new intercellular signaling mechanism, the properties of homeodomain third helices have led to development of the Penetratin family of peptidic vectors. Among the many applications of Penetratins, two will be described. First, it will be shown that the internalization of homeodomains in a gene trap library of ES cells can lead to the identification of homeoprotein transcriptional targets. This procedure demonstrated that the *BPAG1* locus, coding for proteins that bind intermediate filaments in the nervous system and in the skin, is a target of Engrailed and of several other homeoproteins. This locus has been implicated in two pathologies, thus adding to the list of homeoprotein targets of physiopathological interest. A second application has been to internalize into nerve cells several parts of the amyloid precursor protein cytoplasmic tail and to investigate the functional properties of these domains in vivo and in vitro.

The function of several genes and proteins is unknown or only partially known. This is obviously the case for orphan molecules but also for molecules identified on the basis of mutations, or functional studies. In addition, a survey of the literature indicates that many – if not all – genes serve multiple functions. In this period of rapid gene and sequence identification, assigning functions to genes or sequences is thus of primary interest. To that end, gain and loss of function experiments have been developed. In vitro, they are based on the infection or transfection of wild type or mutated genes. These approaches that do not touch the germ line have recently been extended to in vivo models, primarily through electroporation technologies. In vivo, the main access to functional studies is through the production of transgenic animals with wild type or modified genes or genome fragments. These approaches are indeed very useful, but they have several drawbacks. In vitro transfection or infection is not quantitative

Research and Perspectives in Alzheimer's Diseases
Beyreuther/Christen/Masters (Eds.)
Neurodegenerative Disorders
© Springer-Verlag Berlin Heidelberg 2001

and can be difficult to apply to fragile cells, in particular neurons and lympho-cytes. In addition, the levels of expression are poorly controlled, leading to arti-factual situations of relative physiological value. The in vivo experiments, in par-ticular the knock-in approaches that preserve the physiological transcriptional controls, are more appropriate for functional studies but are time consuming and cannot be used for an easy and rapid first screen. In recent years, we have made an observation that may help to solve some of the problems encountered in gene functional studies. A short peptide, corresponding to the third helix of the home-odomain of the Antennapedia transcription factors, was identified that can be used to internalize proteins, antisense and other hydrophilic reagents into live cells. The present paper describes several applications of this new technology in gain/loss of function experiments.

Homeoproteins Regulate Neuronal Morphology

Homeoproteins define a class of transcription factors involved in multiple biolog-ical processes, primarily but not only during development (Krumlauf 1994), and characterized by their DNA-binding domain: the homeodomain. The homeodo-main is highly conserved across homeoproteins and across species (Gehring et al. 1994). Its sequence, 60 amino acids long, is composed of three α-helices. Home-oproteins are expressed in all tissues, including the central nervous system, where they are responsible for the early differentiation of large morphological domains. For example, the genetic deletion of *Engrailed-1*, a homeogene expressed in the midbrain/hindbrain, leads to an early total disappearance of this territory in the mouse (Joyner 1996). In addition to their early patterning func-tion, homeogenes have a role late in development. Indeed, mutations in homeo-genes can modify the specificity of axonal pathways (see, for example, Le Mouel-lic et al. 1992) and synapse formation (see, for example, Tiret et al. 1998). Accord-ingly, they are expressed throughout development and, in fact, in adulthood, sug-gesting a possible role in the morphological plasticity that characterizes the developing and adult vertebrate nervous system.

In an attempt to investigate the function of homeoproteins in neurite out-growth, we adapted a protocol aimed at antagonizing the activity of endogenous homeoproteins through the mechanical internalization of homeodomains into post-mitotic neurons (Fig. 1). During the latter experiments we accidently dis-covered that the homeodomain of Antennapedia, a *Drosophila* transcription fac-tor, is internalized by cells in culture (Joliot et al. 1991a). This capture takes place at 4°C and at 37°C and does not depend on classical endocytosis, and the homeo-domain is directly addressed to the cytoplasm and eventually to the nucleus of the cells. We then generated several point mutations in the homeodomain (Bloch-Gallego et al. 1993; Le Roux et al. 1993). As summarized in Fig. 1, some mutants, still internalized but deprived of their specific DNA-binding properties, did not stimulate neurite elongation, suggesting a transcriptional effect of the homeodomain (Le Roux et al. 1995) and demonstrating a role for homeoproteins in the morphological differentiation of post-mitotic neurons.

A

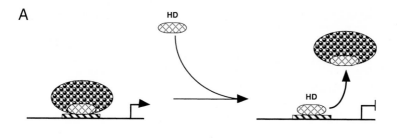

B

AntpHD mutant	Peptidic sequence	Entry	in vitro DNA binding	Biological effect
AntpHD	--AHALCLTE*RQIKIWFQN*RRMKWKKEN	+++	+++	+++
AntpHD50A	--A**Y**ALCLTE*RQIKIWF***A**N*RRMKWKKEN	+++	+	-
AntpHD40P2	--AHALC**PPE***RQIKIWFQN*RRMKWKKEN	++	-	-
AntpHD48S	--AHALCLTE*RQIKI*--**S**NRRMKWKKEN	-	-	-

Fig. 1. Internalized homeodomains compete with homeoproteins. A. The homeodomain (HD), following internalization by live cells, gains access to the nucleus and competes with endogenous homeoproteins for cognate binding sites. B. Description of the mutations affecting the third helix (italics), the second helix and the turn between helices 2 and 3

Cell-Permeable Peptides

An interesting modification was the removal of two amino acids, a tryptophan and a phenylalanine, present in positions 48 and 49 of the homeodomain, i.e. within the third helix (Le Roux et al. 1995). Because this mutated homeodomain was not capable of translocating across biological membranes, the third helix might, in part, be responsible for the unexpected translocating properties of the entire homeodomain. The third helix was synthesized and its internalization by live cells in culture was followed. This peptide of 16 residues, which corresponds almost exactly to the third helix of the homeodomain (amino acids 43 to 58 of the homeodomain), was internalized with an efficiency comparable to that of the entire homeodomain (Derossi et al. 1994). Shorter (15 amino acid-long) peptides failed to translocate across the membrane, and it was proposed that the 16 amino acid-long polypeptide is necessary and sufficient for internalization. This peptide was baptized Penetratin-1 and used as a matrix for the synthesis of several other peptides with translocating properties that now define the Penetratin family (Fig. 2; reviewed in Derossi et al. 1998). All of these peptides are addressed directly to the cytoplasm of cells from which they can be retrieved without apparent degradation. As opposed to Penetratin-1, some variants, in particular those with one or three prolines, do not travel from the cytoplasm to the nucleus and may, therefore, be used for a more specific cytoplasmic targeting. Transloca-

Synthesized peptide		Subcellular localization	
Name	Sequence	Cytoplasm	Nucleus
43-58	RQIKIWFQNRRMKWKK	+++	+++
58-43	**KKWIKMRRNQFWIKIQR**	+++	+++
D43-58	**RQIKIWFQNRRMKWKK**	+++	+++
Pro50	RQIKIWFPNRRMKWKK	+++	+/-
3Pro	RQPKIWFPNRRKPWKK	+++	+/-
Met-Arg	RQIKIWFQNMRRKWKK	+++	+/-
7Arg	RQIRIWFQNRRMRWRR	+++	+++
W/R	**RRWRRWWRRWWRRWRR**	+++	+++
41-55	TERQIKIWFQNRRMK	-	-
46-60	KIWFQNRRMKWKKEN	-	-

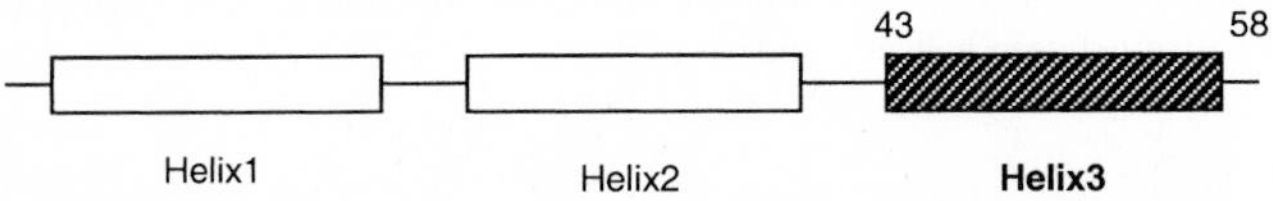

Fig. 2. Modifications of the third helix with the properties of modified peptides. The figure is self-explanatory except for D43–58 (third helix composed of D amino acids)

tion across the plasma membrane is not concentration-dependent (at least between 10 pM and 100 μM) and toxicity is rare below 10 μM.

Mechanism of Translocation

To investigate the presence of a chiral receptor, two peptides were synthesized: a peptide in which the order of the amino acids has been reversed (58–43 instead of 43–58) and a peptide entirely composed of D-enantiomers (Fig. 2). The two peptides are internalized, precluding – almost certainly – the presence of a chiral receptor. This might explain why Penetratins translocate across the plasma membrane of all cell types tested so far, even though translocation can be modulated by the presence of highly charged macromolecules, for example, polysialic acid (Joliot et al. 1991b). Penetratin-1 is poorly structured in water but adopts a helical structure in hydrophobic environments (Derossi et al. 1994). Braking helicity by introducing one or three proline residues within the sequence did not block the internalization, precluding that translocation across the plasma membrane requires a helical conformation (Derossi et al. 1994).

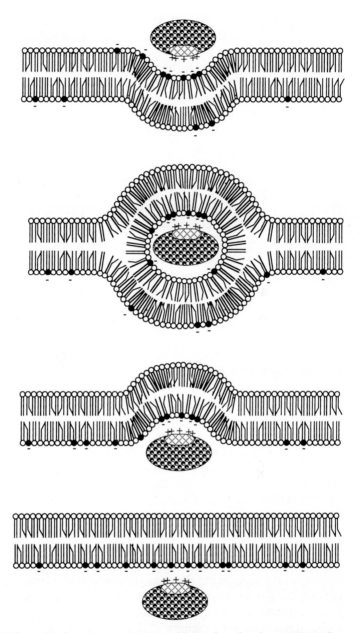

Fig. 3. The inverted micelle model of translocation. In this model, based on phosphorus NMR studies, the protein binds to negative charges exposed at the surface of the membrane. Tryptophan in position 48 of the homeodomain helps in destabilizing the bilayer, provoking the formation of an inverted micelle in which the protein is entrapped. The inverted micelle will then re-open either outside or inside, thus releasing part of the captured protein into the cytoplasm

A peptide in which the two tryptophan residues (positions 43 and 58) have been replaced by two phenylalanins, and is not internalized (Derossi et al. 1994), was used to compare its interactions with brain phospholipids to those of Penetratin-1. It was observed by ^{31}P-NMR that Penetratin-1, but not the non-internalized variant, provokes the formation of inverted micelles (Berlose et al. 1996; unpublished observations). In support of the inverted micelle hypothesis, fluorescence and ^{1}H-NMR spectroscopies demonstrated that the conformational flexibility of the peptide backbones is compatible with their adaptation to the concave surface of SDS micelles and to the convex surface of a reverse micelle (Berlose et al. 1996). On this basis, a model of internalization was proposed in which the peptides, localized in the reverse micelle, travel across the plasma membrane within a hydrophilic pocket (Fig. 3). Direct association of the peptides with the membrane probably involves electrostatic interactions which, in SDS micelles, can be due to SDS itself and, in natural membranes, may require the presence of charged phospholipids, gangliosides, glycosaminoglycans or polysialic acid. It is noteworthy in this context that polysialic acid increases the rate of internalization of the Antennapedia homeodomain by a factor of four (Joliot et al. 1991b). A similar model based on inverted micelles formation has been proposed for the translocation of apocytochrome c into mitochondria (deKruijff et al. 1985).

Biotechnological Applications

In the inverted micelle model, the peptides remain inside a hydrophilic environment. It was thus plausible that hydrophilic molecules linked to Penetratins might be internalized by live cells. This was proven to be the case and several applications of this finding have been developed and recently reviewed (Derossi et al. 1998). A large number of oligonucleotides, peptides and phosphopeptides have been internalized in vitro. For example, specific phosphopeptides interacting with the SH2 domains of Grb2 or PLCγ were developed to block the signaling pathways of FGF or EGF receptors in a highly specific manner (Calvet et al. 1998; Hall et al. 1996; Hildt and Oess 1999; Williams et al. 1997). Another interesting application is the in vitro blockage of the interaction between p16 and cyclin-dependent kinases (Fahraeus et al. 1996).

It is worth mentioning that penetratins have also been used in vivo. One in vivo application has been the induction of the T-cell response by specific antigenic peptides linked to the homeodomain of Antennapedia and internalized by antigen-presenting cells (Schutze-Redelmeier et al. 1996). The principle of the experiment is to address the epitope into the cell cytoplasm to allow its presentation in the MHC-I context. A second in vivo application is with peptide nucleic acids (PNAs). PNAs are oligonucleotides in which the sugar-phosphate backbone has been replaced with a neutral peptide backbone. This modification confers to the molecules the specificity of antisense oligonucleotides and the resistance of peptides. Unfortunately, PNAs are only poorly internalized by live cells, which has limited their use (until now) in vitro and in vivo. Recently, it was shown that the PNAs directed against the type-1 galanin receptor (Gal-R1) and linked to Penetratin-1 are internalized in vivo. Following internalization, they specifically down-regulate the synthesis of Gal-R1 and the physiological activity of galanin (Pooga et al. 1998). Finally, the third helix of the Antennapedia homeodomain has been used to transport doxorubicin across the blood-brain barrier (Rousselle et al. 2000) and to internalize a short peptidic sequence that is part of the cytoplasmic tail of the amyloid precursor protein, resulting in the induction of caspase 3 and in neuronal apoptosis (unpublished results).

Search for Homeoprotein Target Genes

Compared with the amount of information that has accumulated regarding the role and the importance of homeogenes, our understanding of their precise mode of action at the cellular level is poor. This is primarily due to the small number of homeoprotein targets, direct or indirect, identified so far. It thus obvious that the physiological significance of homeoprotein expression requires that homeoprotein targets be identified. Since translocated homeodomains retain transcriptional capabilities (Le Roux et al. 1995), a screen was developed that is based on an induction gene trap approach (Forrester et al. 1996; Hill and Wurst 1993). In this screen (Fig. 4), we have internalized the homeodomain of *Engrailed-2* (EnHD) into ES cells. The use of EnHD as a chemical inducer has allowed us to identify several candidate target loci. For example, it was shown that *Bullous Pemphigoid Antigen 1* (BPAG1) is an Engrailed transcriptional target locus (Mainguy et al. 1999, Mainguy et al. 2000). In fact, EnHD binds to promoters also regulated by homeoproteins other than En2, including En1, and, most likely, by several homeoproteins of the Q50 family that, as is the case for Engrailed, present a glutamine in position 50 of their homeodomain (Biggin and McGinnis 1997; Carr and Biggin 1999). Therefore, this strategy can also lead to the identification and simultaneous mutation of genes regulated by homeoproteins other than Engrailed.

Conclusion: The Paracrine Hypothesis

The third helix is a highly conserved structure among homeodomains (Gehring et al. 1994), suggesting that translocation might represent an intrinsic property of many homeodomains and homeoproteins. This was verified for the homeodomains of Fushi tarazu, Engrailed, Hoxa-5 and Hoxc-8 and for full-length homeoproteins. In particular, it was shown that Hoxa-5 is internalized by neurons in culture and targeted to their nuclei (Chatelin et al. 1996). Internalization occurs at very low concentrations (in the pM range), is not inhibited at 4°C and requires the presence of the homeodomain. It is thus likely that the mechanisms responsible for Hoxa-5 internalization are similar to those used by Penetratins. In the context that homeoproteins may transfer between cells (Prochiantz and Théodore 1995), it was shown that Engrailed is both secreted and internalized; the mechanisms involved in this non-conventional secretion have been partly elucidated (Joliot et al. 1997, 1998; Maizel et al. 1999).

In vivo transfer of homeodomain protein to adjacent non-transcribing cells has been reported in plants. Knotted-1, a plant homeoprotein synthesized in the maize vegetative shoot apex, translocates into one row of L1 epidermal cells that do not express its mRNA (Lucas et al. 1995). This is the first reported case of an in vivo paracrine activity of a recognized homeoprotein. Indeed, this translocation into the L1 cells involves plasmodesmata, a specialized structure found only in plants. It seems possible that animals may have developed other molecular mechanisms to achieve a similar function. In this context it is noteworthy that

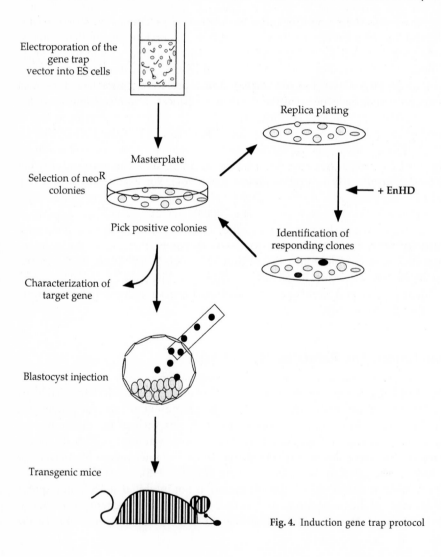

Fig. 4. Induction gene trap protocol

homeoprotein Emx-1 is targetted into nuclei but also into the axons of olfactory neurons (Briata et al. 1996). Because sphingolipid-cholesterol microdomains are primarily addressed to the axonal compartment, the transport of Emx-1 into the axons suggests that the observation reported in Joliot et al. (1997) might apply to other homeoproteins and to specific in vivo situations. We speculate that intercellular homeoprotein transport would provide an interesting mechanism to establish a border between abutting territories or to code for recognition events between cells of different topological origins.

References

Berlose JP, Convert O, Derossi D, Brunissen A, Chassaing G (1996) Conformational and associative behaviours of the third helix of Antennapedia homeodomain in membrane -mimetic environments. Eur J Biochem 242: 372–386

Biggin MD, McGinnis W (1997) Regulation of segmentation and segmental identity by Drosophila homeoproteins: the role of DNA binding in functional activity and specificity. Development 124: 4425–4433

Bloch-Gallego E, Le Roux I, Joliot AH, Volovitch M, Henderson CE, Prochiantz A (1993) Antennapedia homeobox peptide enhances growth and branching of embryonic chicken motoneurons in vitro. J Cell Biol 120: 485–492

Briata P, Di Blas E, Gulisano M, Mallamaci M, Iannone R, Boncinelli E, Corte G (1996) EMX1 homeoprotein is expressed in cell nuclei of the developing cerebral cortex and in the axons of the olfactory sensory neurons. Mech Dev 57: 169–180

Calvet S, Doherty P, Prochiantz A (1998) Identification of a signaling pathway activated specifically in the somatodendritic compartment by a heparan sulfate that regulates dendrite growth. J Neurosci 18: 9751–9765

Carr A, Biggin MD (1999) A comparison of in vivo and in vitro DNA-binding specificities suggests a new model for homeoprotein DNA binding in Drosophila embryo. EMBO J 18: 1598–1608

Chatelin L, Volovitch M, Joliot AH, Perez F, Prochiantz A (1996) Transcription factor Hoxa-5 is taken up by cells in culture and conveyed to their nuclei. Mech Dev 55: 111–117

deKruijff B, Cullis PR, Verkleij AJ, Hope MJ, vanEchteld CJA, Taraschi TF, vanHoogevest P, Killian JA, Rietvel A, vanderSteen ATM (1985) Modulation of lipid polymorphism by lipid-protein interactions. Progress in protein-lipid interactions. Amsterdam, Elsevier Science Publishers BV, 89–142

Derossi D, Joliot AH, Chassaing G, Prochiantz A (1994) The third helix of Antennapedia homeodomain translocates through biological membranes. J Biol Chem 269: 10444–10450

Derossi D, Chassaing G, Prochiantz A (1998) Trojan peptides: the penetratin system for intracellular delivery. Trends Cell Biol 8: 84–87

Fahraeus R, Paramio JM, Ball KL, Lain S, Lane DP (1996) Inhibition of pRb phosphorylation and cell-cycle progresion by a 20-residue peptide derived from p16$^{CDKN2/INK4A}$. Curr Biol 6: 84–91

Forrester LM, Nagy A, Sam M, Watt A, Stevenson L, Bernstein A, Joyner AL, Wurst W (1996) An induction gene trap screen in embryonic stem cells: Identification of genes that respond to retinoic acid in vitro. Proc Natl Acad Sci USA 93: 1677–1682

Gehring WJ, Qian YQ, Billeter M, Furukubo-Tokunaga K, Schier AF, Resendez-Perez D, Affolter M, Otting G, Wüthrich K (1994) Homeodomain-DNA recognition. Cell 78: 211–223

Hall H, Williams EJ, Moore SE, Walsh FS, Prochiantz A, Doherty P (1996) Inhibition of FGF-stimulated phosphatidylinositol hydrolysis and neurite outgrowth by a cell-membrane permeable phosphopeptide. Curr Biol 6: 580–587

Hildt E, Oess S (1999) Identification of Grb2 as a novel binding partner of tumor necrosis factor (TNF) receptor I. J Exp Med 189: 1707–1714

Hill DP, Wurst W (1993) Screening for novel pattern formation genes using trap approaches. Methods Enzymol 225: 664–681

Joliot A, Pernelle C, Deagostini-Bazin H, Prochiantz A (1991a) Antennapedia homeobox peptide regulates neural morphogenesis. Proc Natl Acad Sci USA 88: 1864–1868

Joliot AH, Triller A, Volovitch M, Pernelle C, Prochiantz A (1991b) α-2,8-Polysialic acid is the neuronal surface receptor of Antennapedia homeobox peptide. New Biol 3: 1121–1134

Joliot A, Trembleau A, Raposo G, Calvet S, Volovitch M, Prochiantz A (1997) Association of Engrailed homeoproteins with vesicles presenting caveolae-like properties. Development 124: 1865–1875

Joliot A, Maizel A, Rosenberg D, Trembleau A, Dupas S, Volovitch M, Prochiantz A (1998) Identification of a signal sequence necessary for the unconventional secretion of Engrailed homeoprotein. Curr Biol 8: 856–863

Joyner AL (1996) Engrailed, Wnt and Pax genes regulate midbrain-hindbrain development. Trends Genet 12: 15–20

Krumlauf R (1994) Hox genes in vertebrate development. Cell 78: 191–201

Le Mouellic H, Lallemand Y, Brûlet P (1992) Homeosis in the mouse induced by a null mutation in the Hox-3.1 gene. Cell 69: 251–264

Le Roux I, Joliot AH, Bloch-Gallego E, Prochiantz A, Volovitch M (1993) Neurotrophic activity of the Antennapedia homeodomain depends on its specific DNA-binding properties. Proc Natl Acad Sci USA 90: 9120–9124

Le Roux I, Duharcourt S, Volovitch M, Prochiantz A, Ronchi E (1995) Promoter-specific regulation of genes expression by an exogenously added homeodomain that promotes neurite growth. FEBS Lett 368: 311–314

Lucas WJ, Bouché-Pillon S, Jackson DP, Nguyen L, Baker L, Ding B, Hake S (1995) Selective trafficking of KNOTTED1 homeodomain protein and its mRNA through plasmodesmata. Science 270: 1980–1983

Mainguy G, Erno H, Montesinos ML, Lesaffre B, Wurst W, Volovitch M, Prochiantz A (1999) Regulation of epidermal bullous pemphigoid antigen 1 (BPAG1) synthesis by homeoprotein transcription factors. J Invest Derm 113: 643–650

Mainguy G, Montesinos ML, Lesuffre B, Zevnik B, Karasawa M, Kothary R, Wurst W, Prochiantz A, Volovitch M (2000) An induction gene trap for identifying a homeoprotein-regulated Ions. Nature Biotech 18: 746–749

Maizel A, Bensaude O, Prochiantz A, Joliot AH (1999) A short region of its homeodomain is necessary for Engrailed nuclear export and secretion. Development 126: 3183–3190

Pooga M, Soomets U, Hällbrink, Valkna MA, Saar K, Rezaei K, Kahl U, Hao JX, Xu XJ, Wiesenfeld-Hallin Z, Hökfeld T, Bartfai T, Langel Ü (1998) Cell penetrating PNA constructs down regulate galanin receptor expression and modify pain transmission in vivo. Nature Biotech 16: 857–861

Prochiantz A, Théodore L (1995) Nuclear/growth factors. BioEssays 17: 39–45

Rousselle C, Clair P, Lefauconnier JM, Kackzorek M, Scherrmann JM, Temsamani J (2000) New advances in the transport of doxorubicin through the blood brain barrier by a peptide vector mediated strategy. Mol Pharmacol 57: 679–686

Schutze-Redelmeier MP, Gourmier H, Garcia-Pons F, Moussa M, Joliot AH, Volovitch M, Prochiantz A, Lemonnier F (1996) Introduction of exogenous antigen into the MHC Class I processing and presentation pathway by *Drosophila* Antennapedia homeodomain primes cytotoxic T cells in vivo. J Immunol 157: 650–655

Tiret L, Le Mouellic H, Maury M, Brûlet P (1998) Increased apoptosis of motoneurons and altered somatotopic maps in the brachial spinal cord of Hoxc-8-deficient mice. Development 125: 279–291

Williams EJ, Dunican DJ, Green PJ, Howell FV, Derossi D, Walsh FS, Doherty P (1997) Selective inhibition of growth factor-stimulated mitogenesis by a cell-permeable Grb2-binding peptide. J Biol Chem 272: 22349-22354

Overexpression of APPL, a Drosophila APP Homologue, Compromises Microtubule Associated Axonal Transport and Promotes Synapse Formation

L. Torroja, M. Packard, V. Budnik, and K. White

Introduction

An understanding of the mechanisms that underlie the cellular basis of neural degenerative diseases is important for treating preventing and ultimately curing the disease. In the last decade it has become increasingly clear that cellular processes, be they involved in cell generation, differentiation, maintenance or death, are highly conserved at the molecular level in the animal kingdom. Model genetic systems like the fruit fly Drosophila melanogaster and the nematode Caenorhabditis elegans have contributed enormously to our understanding of the transcriptional and signaling cascades that are important in these processes. Moreover, vertebrate studies of the molecules first identified in invertebrates have reinforced the concept of evolutionary conservation. The ease of doing genetic-molecular studies and the availability of the near complete genome sequence for these genetic model organisms (Adams et al. 2000; The C. elegans Sequencing Consortium 1998) have also made them attractive for the study of neurodegenerative diseases (Anderton 1999; Min and Benzer 1997, 1999; Warrick et al. 1998). For example, a recent impressive case of such a study in Drosophila has shown that polyglutamine-mediated neurodegeneration can be suppressed in vivo by expression of the molecular chaperone HSP70 (Warrick et al. 1999). Therefore, Drosophila can be used not only to understand the mechanisms underlying neurodegenerative diseases but also to search for treatments. Our lab has applied a genetic approach to studying Alzheimer's disease (AD), one of the most common, yet poorly understood neurodegenerative diseases. Using a gain-of-function approach, we have shown that APPL can promote synapse formation at the Drosophila neuromuscular junction (Torroja et al. 1999b), and that overexpression of APPL and Tau disrupts axonal vesicle transport (Torroja et al. 1999a).

APPL is a Drosophila Member of the APP Family of Proteins

Our focus is to elucidate the function of APPL, a Drosophila protein that belongs to the APP family of proteins. APP is the precursor from which the amyloid β-peptide ($A\beta$), the major protein component found in the senile amyloid plaques typical of Alzheimer's diseased brains, is produced. APP is a membrane spanning protein with a large ectodomain and a short cytoplasmic domain (Kang et al.

Research and Perspectives in Alzheimer's Diseases
Beyreuther/Christen/Masters (Eds.)
Neurodegenerative Disorders
© Springer-Verlag Berlin Heidelberg 2001

a.

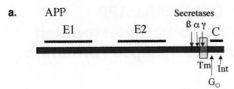

b.   Domain structure of mutant APPL proteins

APPL$^+$	E1	E2	α	Tm	C
APPL^{sd}	E1	E2	$\Delta \alpha$	Tm	C
APPL$^{sd}\Delta$C	E1	E2	$\Delta \alpha$	Tm	ΔC
APPL$^{sd}\Delta$Ci	E1	E2	$\Delta \alpha$	Tm	ΔInt
APPL$^{sd}\Delta$Cg	E1	E2	$\Delta \alpha$	Tm	ΔG$_O$
APPL$^{sd}\Delta$E1	ΔE1	E2	$\Delta \alpha$	Tm	ΔE1
APPL$^{sd}\Delta$E2	E1	ΔE2	$\Delta \alpha$	Tm	ΔE2

Fig. 1.a Amyloid Precursor Protein (APP) schematic showing the Aβ peptide in brick color and the sites of β, α, and γ cleavage. The domains conserved in Drosophila APPL are shown as black bars: E1, E2, and C are the regions of high amino acid homology. Also noted are: α-secretase-like cleavage site (α), transmembrane domain (Tm), putative G$_O$ binding site (G$_O$), and internalization signal (Int.) **b** Table indicating the domain structure of the APPL wild type and mutant proteins used in the gain-of-function analysis. The domains deleted are shown in red

1987). The Drosophila Appl gene was reported two years after the human APP gene (Rosen et al. 1989). APPL is similar to APP throughout the molecule, with high homology in the ectodomain regions E1 and E2 and in the highly conserved cytoplasmic domain (see Fig. 1). Unlike APP, APPL is expressed only in neurons and encodes only one isoform, making it most similar to the neural enriched APP^{695} spliced form. Another important difference is that APPL and two other vertebrate APP family members, APLP1 and APLP2, do not show similarity to APP in the Aβ-peptide region (Rosen et al. 1989; Wasco et al. 1992, 1993).

Understanding APP processing is crucial to revealing the mechanisms responsible for the formation of the Aβ peptide and amyloid plaques. It is commonly accepted that all members of the APP family undergo proteolytic cleavage at the α-secretase site, which releases the large soluble ectodomain (Luo et al. 1990; Sisodia et al. 1994; Slunt et al. 1994; Weidemann et al. 1989). The release of the Aβ-peptide, however, requires action by two distinct secretases: the β-secretase that cleaves in the ectodomain, and the γ-secretase that cleaves within the transmembrane domain (Fig. 1). In the past several years, two human presenilin proteins encoded by distinct genes, presenilin-1 and presenilin-2, have been implicated in AD (reviewed in Selkoe 1998). About 10 % of the early onset Familial Alzheimer's disease (FAD) cases show mutations in either of the presenilin genes. Current data demonstrate that Presenilin 1 is required for the γ-cleavage of APP (De Strooper et al. 1998; Wolfe et al. 1999). Interestingly, presenilins are also involved in the processing of Notch (De Strooper et al. 1999; Struhl and Greenwald 1999; Ye et al. 1999), a transmembrane receptor that occupies a pivotal position in signaling cascades used in developmental decisions in invertebrates as well as vertebrates. Like APP γ-cleavage, Notch cleavage occurs within the transmembrane domain. Importantly, in Drosophila, Notch cleavage has been shown to be essential for transduction of Notch signaling, because loss-of-function mutations of the presenilin gene lead to defective Notch cleavage and to Notch-like phenotypes (Guo et al. 1999; Struhl and Greenwald 1999; Ye et al.

1999). These studies on Notch, along with data published on APP expressed in flies (Fossgreen et al. 1998), establish that both α-secretase and γ-secretase-like activities are present in flies. At present there is no published evidence for a β-secretase-like activity in Drosophila. Thus, the two major known factors in AD, APP and presenilins, are conserved from flies to humans not only in their basic structure but also in important biological properties.

The Function of APP and APPL Proteins

A clear understanding of the normal function of APP has eluded researchers for a long time. Traditionally, analysis of loss-of-function mutations is a well-tried method to garner information on the function of a gene. Neither Appl loss-of-function mutant flies nor APP knockout mice have been particularly helpful, as they are viable animals with no apparent developmental problems, which suggests that these proteins in themselves are not vital to the development of the organism. Although they appear morphologically normal, APPL-deficient flies do exhibit subtle behavioral defects, including deficits in fast phototaxis and poor reactivity (Luo et al. 1992). The APP knockout mice also show impaired learning and reduced locomotor activity, accompanied by increased gliosis (Muller et al. 1994; Zheng et al. 1995). Two pieces of evidence suggest that the APP-family proteins could have overlapping functional capability. First, the phototaxis deficit of APPL-deficient flies was ameliorated by either APPL-expressing transgenes or by APP-expressing transgenes (Luo et al. 1992). Second whereas mice deficient in APP or APLP2 genes are morphologically normal, APP/APLP2 double knockouts show early postnatal lethality (von Koch et al. 1997), suggestive of an overlap in functions.

The loss-of-function mutant analyses demonstrate that APPL and APP, although conserved in the vertebrate and invertebrate phyla, are not essential cellular proteins. Nevertheless, these proteins are present in neurons from the inception of axonogenesis and are highly enriched in areas of synaptogenesis. Collectively, these properties hint at a regulatory role for APPL and APP in neurons. Finding the cellular processes that are regulated by APP has not been an easy task. One difficulty is the complex processing of APP, with different activities described for the different protein products, i.e., secreted, transmembrane and cleaved peptides (for review see Mattson 1997). In addition, APP contains numerous domains that can interact with specific molecules, like extracellular proteins, cytoskeletal proteins, and intracellular signaling molecules (reviewed in Russo et al. 1998). A receptor role for APP was originally proposed based on its transmembrane receptor-like structure. Recently, this notion has gathered force. The cytoplasmic domain of different members of the APP family can bind the same intracellular proteins (Bressler et al. 1996; Duilio et al. 1998; Homayouni et al. 1999), many of which act in intracellular signaling cascades, suggesting conserved intracellular signaling mechanisms. Another line of investigation provides evidence that APP is involved in G_0 protein mediated signaling (Nishimoto et al.

1993; Okamoto et al. 1996; Yamatsuji et al. 1996). It has also been proposed that APP interacts indirectly, via its cytoplasmic domain, with ApoE receptors (Trommsdorff et al. 1998). However, the processes where APP receptor activity is required remain unknown.

Our recent work has used gain-of-function strategies, because gain-of-function analysis has often provided functional insights for genes with overlapping, redundant, or subtle regulatory functions. In the fruit fly, the UAS/GAL4 system provides a powerful method to express a cloned gene of choice in a tissue-specific manner (Brand and Perrimon 1993). In this system, the yeast transcrip-

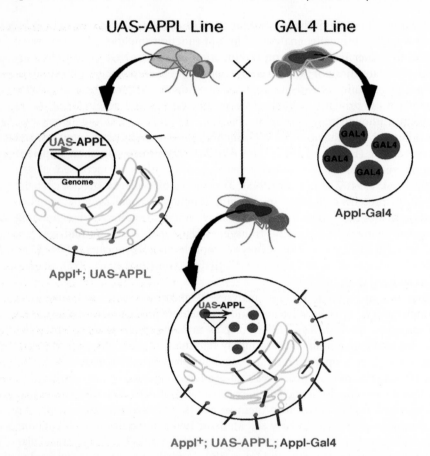

Fig. 2. Scheme of overexpression of APPL using the UAS/GAL4 system. Flies carrying a UAS-APPL transgene are mated to flies carrying the APPL-GAL4 transgene. Progeny flies carrying both transgenes, APPL-GAL4 and UAS-APPL, overexpress APPL. The APPL-GAL4 transgene expresses the yeast transcriptional activator under the endogenous Appl promoter in neurons, and drives the expression of UAS-APPL. The production of transmembrane APPL protein, blue rod with red cytoplasmic domain, is depicted in the neurons with UAS-APPL transgene and with UAS-APPL and APPL-GAL4 transgenes. In the former (Appl+; UAS-APPL), the protein is produced only from the endogenous gene, whereas in the latter (Appl+; UAS-APPL; UAS-GAL4), protein is also made from the UAS-APPL transgene (Adapted from Toba et al. 1999)

tional activator GAL4 directs transcription of the gene of choice, which has been placed under the transcriptional control of Upstream Activating Sequence (UAS). For example, to overexpress APPL in neurons, flies carrying the UAS-APPL transgene, in which Appl cDNA is located downstream of the UAS promoter, are crossed to flies carrying the Appl-GAL4 transgene, which expresses GAL4 in most neurons (Fig. 2). Flies carrying both of these transgenes express APPL at levels about three-fold higher than the endogenous gene at 30°C (Torroja et al. 1999b).

APPL Promotes Synapse Formation at the Neuromuscular Junction

The behavioral deficits seen in loss-of-function Appl flies and APP-knockout mice suggest neural dysfunction. However, mice overexpressing APP have been reported to display either increased synapse density (Mucke et al. 1994) or Alzheimer's-like pathology without synaptotrophic effects (Higgins et al. 1993; Masliah et al. 1995). The lack of a good quantifiable system could explain these contradictory results. Therefore, we analyzed the effect of APPL-overexpression on synapse formation in Drosophila (Torroja et al. 1999b). Below we summarize the results from this study.

The Drosophila larval neuromuscular junction (NMJ) provides an ideal system to analyze effects on synapse formation, as it has a stereotypic and easily quantifiable morphology. For quantification, a specific NMJ is assessed for each case by counting synaptic contacts on a specific muscle fiber. Larvae with a deletion in the Appl gene have NMJs of normal appearance, but show a small reduction in the number of synaptic boutons. These results imply that APPL is not necessary for the formation of synaptic contacts, but it may regulate this process in response to specific signals. We reasoned that constitutive overexpression of APPL at the NMJ could enhance APPL synaptic activity, thus providing an in vivo assay for studying its mechanism of action. Overexpression was achieved by expressing APPL in motoneurons using the UAS/GAL4 system. Overexpression of wildtype APPL increased the number of synaptic boutons by almost three-fold. This increase affected boutons of normal appearance (mature), as well as smaller boutons (immature or satellite) that resembled early developmental forms of the synaptic bouton (Zito et al. 1999; see Fig. 3). A similar, but weaker effect was observed when human APP695 was expressed in the same cells. These results suggest that APPL has synapse-promoting activity.

To determine which regions of APPL protein are important for synapse formation, a panel of transgenic lines expressing the mutant APPL proteins, shown in Figure 1b, was tested with the same in vivo assay. These constructs expressing mutated APPL proteins were made in a parent construct that encoded a secretion-defective APPL form (APPL^{sd}) lacking the cleavage sequence. Similar to wild type APPL, overexpression of the secretion-defective APPL, form increased the number of synaptic boutons by almost three-fold, also affecting both mature and immature boutons (Fig. 3). Interestingly, the cytoplasmic domain was essen-

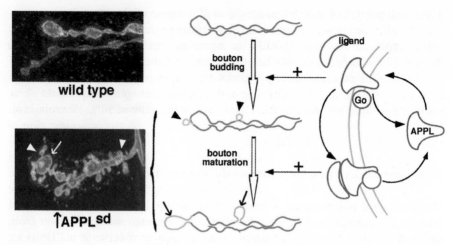

Fig. 3. Synapse promotion by APPL-overexpression. Left panel shows comparable neuromuscular junctions of wild type and APPL-overexpressing flies. Note the presence of small, immature boutons when APPL is overexpressed (arrowheads). The middle panel shows the steps in bouton formation (based on Zito et al. 1999). Arrowheads indicate newly sprouted immature boutons, and arrows indicate their mature state. The model suggests that surface APPL promotes budding of immature boutons, and activation of the APPL receptor induces further maturation. When APPL is overexpressed, both bouton forms are increased (left panel). (Figure summarizes data from Torroja et al. 1999)

tial for this effect, because deletion of the entire cytoplasmic domain completely abolished the increase in bouton number. Thus, APPL is able to modulate the number of synaptic contacts at the NMJ. This finding is consistent with the behavioral defects observed in Appl-null mutants and suggests that APPL may be involved in synaptic plasticity.

Further analysis demonstrated that specific domains of APPL could be distinguished based on their ability to promote formation of either mature or immature boutons. APPL forms deficient for the extracellular domains E1 or E2, or the putative Go-binding site, increased only the number of immature boutons and did not affect mature boutons. Conversely, an APPL form missing the NPTY internalization signal was able to increase the number of mature boutons, with very little effect on the number of immature boutons. These results indicate that APPL acts as a receptor, and that two aspects of the biology of APPL protein modulate its function: the turnover of APPL between the cell surface and cytoplasmic compartments, and the activation of APPL-associated receptor function.

Several Drosophila mutants have been shown to affect the number of synaptic boutons at the NMJ. One class comprises mutants that increase neuronal activity as a result of deficiencies in potassium channel subunits and display NMJs with an increased number of synaptic boutons. For this reason, and because neuronal activity has been shown to regulate APP metabolism in mammalian systems (Allinquant et al. 1994; Nitsch et al. 1994), we investigated how these hyperactive mutations (eag Sh) affect the synaptogenic activity of APPL.

Surprisingly, expressing APPL in a hyperactive neuron was equivalent to expressing the APPL form without the internalization sequence: only the number of mature boutons increased. Moreover, at the hyperactive NMJ, APPL protein appeared in closer association with the plasma membrane.

APPL appears to affect two aspects of bouton formation: budding of an immature bouton and its subsequent maturation. A model explaining these results proposes that translocation of APPL to the plasma membrane induces budding of an immature bouton, but bouton maturation requires activation of the APPL receptor (Fig. 3). Consequently, if APPL is internalized before activation, further development of the bouton is aborted. Any mutation that impedes APPL-receptor activation (like the domain deletions that block ligand-binding or Go binding) will prevent bouton maturation, resulting in an increased number of immature, but not mature, boutons. On the other hand, manipulations that increase the presence of APPL on the plasma membrane (like reduced internalization through the NPTY deletion or increased neuronal activity) will enhance the probability of APPL-receptor activation, and thus promote bouton maturation and increase the number of mature boutons. The use of a genetic-amenable system like Drosophila will enable further testing of this model.

APPL and Tau Overexpression Disrupt Axonal Transport

The other anatomical hallmark of Alzheimer's brains is the presence of intracellular neurofibrillary tangles that have a high concentration of the microtubule-associated axonal protein Tau. An intriguing question is what connections (if any) exist between the formation of the extracellular βA-enriched senile plaques and the intracellular Tau-enriched tangles. At present, there are no direct answers to this question. Nevertheless, APP immunoreactivity has been detected in the neurofibrillary tangles, and direct interaction between APP and Tau and the induction of Tau phosphorylation by APP and Aβ peptide have been reported (Greenberg et al. 1994; Islam and Levy 1997; Smith et al. 1995). Although there are proteins with structure and properties analogous to mammalian Tau in Drosophila (Wandosell and Avila 1987), the Drosophila tau gene has yet to be cloned. However, a bovine Tau isoform containing four microtubule-binding repeats was able to induce the formation of microtubule bundles in vivo, when expressed in Drosophila ring gland cells using the UAS/Gal4 system (Fig. 4a). Therefore, because Tau appears to have the same biological activity in Drosophila cells and in mammalian cells, we used this UAS-Tau transgene to express Tau in neurons and study its interaction with APPL (Torroja et al. 1999a). The results of this study are summarized below.

Pan-neural co-expression of APPL and Tau resulted in decreased viability when compared with overexpression of either APPL or Tau alone. The eclosing adults exhibited an "infantile" phenotype, where wing expansion and cuticle hardening does not occur; this phenotype has been previously associated with neuroendocrine insufficiency (McNabb et al. 1997). Since Tau is known to regu-

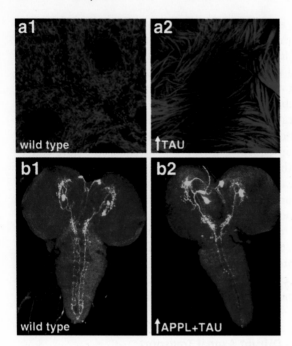

Fig. 4. Effects of APPL and Tau overexpression. **a** Tubulin-immunoreactivity in wild type (**a1**) and Tau-overexpressing (**a2**) large cells of the ring gland. Note the formation of microtubule bundles induced by Tau (**a2**). **b** Visualization of neuropeptide immunoreactivity in the larval central nervous system. An antibody against Drosophila myosuppressin (Nichols et al. 1997) was used to visualize the cell bodies (in upper brain lobes) and processes (in lower ventral ganglion) in wild type (**b1**) and APPL- and Tau-overexpressing (**b2**) central nervous systems. Note that the signal in the distal processes away from the cell body is considerably weaker in b2. (Fig. 4b adapted from Torroja et al. 1999a)

late microtubule-based movement (Ebneth et al. 1998; Sato-Harada et al. 1996), the underlying cellular dysfunction could be related to vesicle transport and/or secretion. To analyze if vesicle transport was affected, synaptic proteins were revealed immunocytochemically. In Drosophila larvae, synaptic proteins concentrate in the NMJ synaptic boutons and are barely detected in motor axons. However, when either APPL or Tau was expressed in motor neurons, these synaptic proteins were retained in the axons, indicating defects in their axonal transport. Studies of neuropeptide immunoreactivity revealed that transport of neuropeptide-containing dense core vesicles was also disrupted in neurons expressing both APPL and Tau, because they displayed a dramatic decrease of immunoreactive signal in the distal processes, whereas proximal signal remained the same (Fig. 4b).

The axonal retention of synaptic proteins is a phenotype also caused by mutations in the Kinesin heavy chain (Khc) gene, a motor for anterograde vesicle trafficking (Hurd and Saxton 1996). Therefore, we tested if a 50 % reduction in Khc protein would enhance APPL-overexpression-associated phenotypes. Indeed, pan-neural APPL overexpression in a sensitized genetic background with 50 % of normal Khc function led to a highly penetrant infantile phenotype. Moreover, APPL overexpression and mutations in Khc showed phenotypic similarities: they both suppressed the uncontrolled leg-shaking behavior of hyperactive Shaker mutations (Hurd et al. 1996; Torroja et al. 1999a).

Despite some similarities, there are clear phenotypic differences between APPL overexpression and Khc mutations: 1) Khc null alleles are lethal (Saxton et al. 1991), whereas a three-fold increase in APPL has little effect on viability; 2) Khc mutants show atrophic larval NMJs (Saxton et al. 1991), contrary to APPL-overexpressing NMJs; and 3) Khc mutant larvae display a tail-flipping phenotype that is not observed in APPL-overexpressing larvae. Thus, although the genetic interaction between APPL and Khc further supports a role for APPL in vesicle trafficking, the subtle and viable APPL-associated phenotypes, both in loss- and gain-of-function mutations, suggest that APPL must function in a regulatory capacity.

Closing Remarks

In this article we have reviewed gain-of-function studies that attempt to reveal the cellular processes that APPL may influence. Neuronal APPL overexpression-associated phenotypes indicate that APPL can modulate synapse formation and affect microtubule-associated trafficking. The synapse-promoting activity is best explained on the basis of APPL receptor activation, with the cytoplasmic domain being essential for transducing the signal. Based on human APP-overexpression studies in transgenic Drosophila, Fossgreen et al. (1998) have proposed a signaling role for the APP protein. In this study they showed that overexpression of APP in the wing cells resulted in wing blisters and that the cytoplasmic domain was essential to generate this phenotype. Since wing blisters reflect disruption between the dorsal and ventral epithelial cell layers, they propose that APP may normally function in cell adhesion and signaling. Physiological roles for APPL or APP in synapse modulation or adhesion are not novel, as many previous studies have indicated such functions (reviewed in Mattson 1997). However, they now provide new ways to uncover the signaling pathways used in these processes.

Is there a connection between the two apparently very different processes influenced by APPL overexpression: synapse formation and vesicle trafficking? The synapse-promoting action could not be a mere secondary effect of the organelle transport defects, because Khc mutants show the opposite effect, i.e., reduced number of synaptic boutons (Saxton et al. 1991). Keeping in mind that APPL must act as a regulatory molecule, one attractive hypothesis is that, in response to external signals, APPL induces recruitment of proteins/lipids necessary for synapse formation by regulating the trafficking of their cargo vesicles between different compartments. The transport-related phenotypes could arise from an imbalance in the overall intracellular trafficking network, caused by the elevated level of APPL acting as a regulatory factor. A proposed role of APPL in vesicle trafficking raises questions about the mechanisms underlying AD. Interestingly, the other main factors in AD, the presenilins, have also been implicated in membrane trafficking (Naruse et al. 1998; Nishimura et al. 1999). Therefore, in addition to the effect of presenilin on APP processing, membrane trafficking could be conceived as the point of connection between these two proteins that have been proven to be causal factors in FAD.

Acknowledgments

This paper is based on original findings presented in two articles by Torroja et al. (1999a, b). This work was supported by Funds from the National Institutes of Health. We thank Gakuta Toba for his help with the figures.

References

Adams MD, Celniker SE, Holt RA, Evans CA, Gocayne JD, Amanatides PG, Scherer SE, Li PW, Hoskins RA, Galle RF, George RA, Lewis SE, Richards S, Ashburner M, Henderson SN, Sutton GG, Wortman JR, Yandell MD, Zhang Q, Chen LX, Brandon RC, Rogers Y-HC, Blazjej RG, Champe M, Pfeiffer BD, Wan KH, Doyle C, Baxter EG, Helt G, Nelson CR, Gabor Miklos GL, Abril JF, Agbayani A, An H-J, Andrews-Pfannkoch C, Baldwin D, Ballew RM, Basu A, Bawendale J, Bayraktaroglu L, Beasley EM, Beeson KY, Benos PV, Berman BP, Bhandari D, Bolshakov S, Borkova D, Botchan MR, Bouck J, Brockstein P, Brottier P, Burtis KC, Busam DA; Butler H, Cadieu E, Center A, Chandra I, Cherry JM, Cawley S, Dahlke C, Davenport LB, Davies P, de Pablos B, Delcher A, Deng Z, Deslattes Mays A, Dew I, Dietz SM, Dodson K, Doup LE, Downes M, Dugan-Rocha S, Dunkov BC, Dunn P, Durbin KJ, Evangelista CC, Ferraz C, Ferriera S, Fleischmann W, Fosler C, Gabrielian AE, Garg NS, Gelbart WM, Glasser K, Glodek A, Gong F, Gorrell JH, Gu Z, Guan P, Harris M, Harris NL, Harvey D, Heiman TJ, Hernandez JR, Houck J, Hostin D, Houston KA, Howland TJ, Wei M-H, Ibegwam C, Jalali M, Kalush F, Karpen GH, Ke Z, Kennison JA, Ketchum KA, Kimmel BE, Kodira CD, Kraft C, Kravitz S, Kulp D, Lai Z, Lasko P, Lei Y, Levitsky AA, Li J, Li Z, Liang Y, Lin X, Liu X, Mattei B, McIntosh TC, McLeod MP, McPherson D, Merkulov M, Milshina NV, Mobarry C, Morris J, Moshrefi A, Mount SM, Moy M, Murphy B, Murphy L, Muzny DM, Nelson DL, Nelson DR, Nelson KA, Nixon K, Nusskern DR, Pacleb JM, Palazzolo M, Pittman GS, Pan S, Pollard J, Puri V, Reese MG, Reinert K, Remington K, Saunders RDC, Scheeler F, Shen H, Shue BC, Sidén-Kiamos I, Simpson M, Skupski MP, Smith T, Spier E, Spradling AC, Stapleton M, Strong R, Sun E, Svirskas R, Tector C, Turner R, Venter E, Wang AH, Wang X, Wang Z-Y, Wassarman DA, Weinstock GM, Weissenbach J, Williams SM, Woodage T, Worley KC, Wu D, Yang S, Yao A, Ye J, Yeh R-F, Zaveri JS, Zhan M, Zhang G, Zhao Q, Zheng L, Zheng XH, Zhong FN, Zhong W, Zhou X, Zhu S, Zhu X, Smith HO, Gibbs RA, Myers EW, Rubin GM, Venter JC (2000) The genome sequence of Drosophila melanogaster. Science 287: 2185–2196
Allinquant B, Moya K, Bouillot C, Prochiantz A (1994) Amyloid precursor protein in cortical neurons: coexistence of two pools differentially distributed in axons and dendrites and association with cytoskeleton. J Neurosci 14: 6842–6854
Anderton BH (1999) Alzheimer's disease: clues from flies and worms. Curr Biol 9: R106–109
Brand AH, Perrimon N (1993) Targeted gene expression as a means of altering cell fates and generating dominant phenotypes. Development 118: 401–415
Bressler SL, Gray MD, Sopher BL, Hu Q, Hearn MG, Pham DG, Dinulos MB, Fukuchi K, Sisodia SS, Miller MA, Disteche CM, Martin GM (1996) cDNA cloning and chromosome mapping of the human Fe65 gene: interaction of the conserved cytoplasmic domains of the human beta-amyloid precursor protein and its homologues with the mouse Fe65 protein. Human Mol Genet 5: 1589–1598
The C. elegans Sequencing Consortium (1998) Genome sequence of the nematode C. elegans: a platform for investigating biology. Science 282: 2012–2018
De Strooper B, Saftig P, Craessaerts K, Vanderstichele H, Guhde G, Annaert W, Von Figura K, Van Leuven F (1998) Deficiency of presenilin-1 inhibits the normal cleavage of amyloid precursor protein. Nature 391: 387–390
De Strooper B, Annaert W, Cupers P, Saftig P, Craessaerts K, Mumm JS, Schroeter EH, Schrijvers V, Wolfe MS, Ray WJ, Goate A, Kopan R (1999) A presenilin-1-dependent gamma-secretase-like protease mediates release of Notch intracellular domain. Nature 398: 518–522
Duilio A, Faraonio R, Minopoli G, Zambrano N, Russo T (1998) Fe65L2: a new member of the Fe65 protein family interacting with the intracellular domain of the Alzheimer's beta-amyloid precursor protein. Biochem J 330: 513–519

Ebneth A, Godemann R, Stamer K, Illenberger S, Trinczek B, Mandelkow E (1998) Overexpression of tau protein inhibits kinesin-dependent trafficking of vesicles, mitochondria, and endoplasmic reticulum: implications for Alzheimer's disease. J Cell Biol 143: 777–794

Fossgreen A, Bruckner B, Czech C, Masters CL, Beyreuther K, Paro R (1998) Transgenic Drosophila expressing human amyloid precursor protein show gamma-secretase activity and a blistered-wing phenotype. Proc Natl Acad Sci USA 95: 13703–13708

Greenberg SM, Koo EH, Selkoe DJ, Qiu WQ, Kosik KS (1994) Secreted beta-amyloid precursor protein stimulates mitogen-activated protein kinase and enhances tau phosphorylation. Proc Natl Acad Sci USA 91: 7104–7108

Guo Y, Livne-Bar I, Zhou L, Boulianne GL (1999) Drosophila presenilin is required for neuronal differentiation and affects notch subcellular localization and signaling. J Neurosci 19: 8435–8442

Higgins LS, Catalano R, Quon D, Cordell B (1993) Transgenic mice expressing human beta-APP751, but not mice expressing beta-APP695, display early Alzheimer's disease-like histopathology. Ann NY Acad Sci 695: 224–227

Homayouni R, Rice DS, Sheldon M, Curran T (1999) Disabled-1 binds to the cytoplasmic domain of amyloid precursor-like protein 1. J Neurosci 19: 7507–7515

Hurd DD, Saxton WM (1996) Kinesin mutations cause motor neuron disease phenotypes by disrupting fast axonal transport in Drosophila. Genetics 144: 1075–1085

Hurd DD, Stern M, Saxton WM (1996) Mutation of the axonal transport motor kinesin enhances paralytic and suppresses Shaker in Drosophila. Genetics 142: 195–204

Islam K, Levy E (1997) Carboxyl-terminal fragments of beta-amyloid precursor protein bind to microtubules and the associated protein tau. Am J Pathol 151: 265–271

Kang J, Lemaire HG, Unterbeck A, Salbaum JM, Masters CL, Grzeschik KH, Multhaup G, Beyreuther K, Müller-Hill B (1987) The precursor of Alzheimer's disease amyloid A4 protein resembles a cell-surface receptor. Nature 325: 733–736

Luo L, Martin-Morris LE, White K (1990) Identification, secretion and neural expression of APPL, a Drosophila protein similar to human amyloid protein precursor. J Neurosci 10: 3849–3861

Luo L, Tully T, White K (1992) Human amyloid precursor protein ameliorates the behavioral deficits of flies deleted for Appl gene. Neuron 9: 595–605

Masliah E, Mallory M, Alford N, Mucke L (1995) Abnormal synaptic regeneration in hAPP695 transgenic and ApoE knockout mice. In: Iqbal K MJ, Winblad B, Wisniewsky H (eds) Research advances in Alzheimer's disease and related disorders. J Wiley New York, pp 405–414

Mattson MP (1997) Cellular actions of beta-amyloid precursor protein and its soluble and fibrillogenic derivatives. Physiol Rev 77: 1081–132

McNabb SL, Baker JD, Agapite J, Steller H, Riddiford LM, Truman JW (1997) Disruption of a behavioral sequence by targeted death of peptidergic neurons in Drosophila. Neuron 19: 813–823

Min KT, Benzer S (1997) Spongecake and eggroll: two hereditary diseases in Drosophila resemble patterns of human brain degeneration. Curr Biol 7: 885–888

Min KT, Benzer S (1999) Preventing neurodegeneration in the Drosophila mutant bubblegum. Science 284: 1985–1988

Mucke L, Masliah E, Johnson WB, Ruppe MD, Alford M, Rockenstein EM, Forss-Petter S, Pietropaolo M, Mallory M, Abraham CR (1994) Synaptotrophic effects of human amyloid beta protein precursors in the cortex of transgenic mice. Brain Res 666: 151–167

Muller U, Cristina N, Li ZW, Wolfer DP, Lipp HP, Rulicke T, Brandner S, Aguzzi A, Weissmann C (1994) Behavioral and anatomical deficits in mice homozygous for a modified beta-amyloid precursor protein gene. Cell 79: 755–765

Naruse S, Thinakaran G, Luo JJ, Kusiak JW, Tomita T, Iwatsubo T, Quian X, Ginty DD, Price DL, Borchelt DR, Wong PC, Sisodia SS (1998) Effects of PS1 deficiency on membrane protein trafficking in neurons. Neuron 21: 1213–1221

Nichols R, McCormick J, Lim I (1997) Multiple antigenic peptides designed to structurally related Drosophila peptides. Peptides 18: 41–45

Nishimoto I, Okamoto T, Matsuura Y, Takahashi S, Murayama Y, Ogata E (1993) Alzheimer amyloid protein precursor complexes with brain GTP-binding protein G(o). Nature 362: 75–79

Nishimura M, Yu G, Levesque G, Zhang DM, Ruel L, Chen F, Milman P, Holmes E, Liang Y, Kawarai T, Jo E, Supala A, Rogaeva E, Xu DM, Janus C, Levesque L, Bi Q, Duthie M, Rozmahel R, Mattila K, Lannfelt L, Westaway D, Mount HT, Woodgett J, St George-Hyslop P (1999) Presenilin mutations associated with Alzheimer disease cause defective intracellular trafficking of beta-catenin, a component of the presenilin protein complex. Nature Med 5: 164–169

Nitsch RM, Slack BE, Farber SA, Schulz JG, Deng M, Kim C, Borghesani PR, Korver W, Wurtman RJ, Growdon JH (1994) Regulation of proteolytic processing of the amyloid beta-protein precursor of Alzheimer's disease in transfected cell lines and in brain slices. J Neural Transm Suppl 44: 21–27

Okamoto T, Takeda S, Giambarella U, Murayama Y, Matsui T, Katada T, Matsuura Y, Nishimoto I (1996) Intrinsic signaling function of APP as a novel target of three V642 mutations linked to familial Alzheimer's disease. EMBO J 15: 3769–3777

Rosen RR, Martin-Morris L, Luo L, White K (1989) A Drosophila gene encoding a protein resembling the human β-amyloid protein precursor. Proc Natl Acad Sci USA 86: 2478–2482

Russo T, Faraonio R, Minopoli G, De Candia P, De Renzis S, Zambrano N (1998) Fe65 and the protein network centered around the cytosolic domain of the Alzheimer's beta-amyloid precursor protein. FEBS Lett 434: 1–7

Sato-Harada R, Okabe S, Umeyama T, Kanai Y, Hirokawa N (1996) Microtubule-associated proteins regulate microtubule function as the track for intracellular membrane organelle transports. Cell Struct Funct 21: 283–295

Saxton WM, Hicks J, Goldstein LS, Raff EC (1991) Kinesin heavy chain is essential for viability and neuromuscular functions in Drosophila, but mutants show no defects in mitosis. Cell 64: 1093–1102

Selkoe DJ (1998) The cell biology of beta-amyloid precursor protein and presenilin in Alzheimer's disease. Trends Cell Biol 8: 447–453

Sisodia SS, Slunt HH, Van Koch C, Lo ACY, Thinakaran G (1994) Studies on APP biology: analysis of APP secretion and characterization of an APP homologue, APLP2. In: Masters CL, Beyreuther K, Trillet M and Christen Y (eds) Amyloid Protein Precursor in Development, Aging and Alzheimer's Disease. Springer-Verlag, pp 121–133

Slunt HH, Thinakaran G, Koch CV, Lo AC, Tanzi RE, Sisodia SS (1994) Expression of a ubiquitous, cross-reactive homologue of the mouse beta-amyloid precursor protein (APP). J Biol Chem 269: 2637–2644

Smith MA, Siedlak SL, Richey PL, Mulvihill P, Ghiso J, Frangione B, Tagliavini F, Giaccone G, Bugiani O, Praprotnik D, Kalaria RN, Perry G (1995) Tau protein directly interacts with the amyloid beta-protein precursor: implications of Alzheimer's disease. Nature Med 1: 365–369

Struhl G, Greenwald I (1999) Presenilin is required for activity and nuclear access of Notch in Drosophila. Nature 398: 522–525

Toba G, Ohsako T, Miyata N, Ohtsuka T, Seong KH, Aigaki T (1999) The gene search system. A method for efficient detection and rapid molecular identification of genes in Drosophila melanogaster. Genetics 151: 725–737

Torroja L, Chu H, Kotovsky I, White K (1999a) Neuronal overexpression of APPL, the Drosophila homologue of the amyloid precursor protein (APP), disrupts axonal transport. Curr Biol 9: 489–492

Torroja L, Packard M, Gorczyca M, White K, Budnik V (1999b) The Drosophila beta-amyloid precursor protein homolog promotes synapse differentiation at the neuromuscular junction. J Neurosci 19: 7793–7803

Trommsdorff M, Borg JP, Margolis B, Herz J (1998) Interaction of cytosolic adaptor proteins with neuronal apolipoprotein E receptors and the amyloid precursor protein. J Biol Chem 273: 33556–33560

von Koch CS, Zheng H, Chen H, Trumbauer M, Thinakaran G, van der Ploeg LH, Price DL, Sisodia SS (1997) Generation of APLP2 KO mice and early postnatal lethality in APLP2/APP double KO mice. Neurobiol Aging 18: 661–669

Wandosell F, Avila J (1987) Microtubule-associated proteins present in different developmental stages of Drosophila melanogaster. J Cell Biochem 35: 83–92

Warrick JM, Paulson HL, Gray-Board GL, Bui QT, Fischbeck KH, Pittman RN, Bonini NM (1998) Expanded polyglutamine protein forms nuclear inclusions and causes neural degeneration in Drosophila. Cell 93: 939–949

Warrick JM, Chan HY, Gray-Board GL, Chai Y, Paulson HL, Bonini NM (1999) Suppression of polyglutamine-mediated neurodegeneration in Drosophila by the molecular chaperone HSP70. Nature Genet 23: 425–428

Wasco W, Bupp K, Magendantz M, Gusella JF, Tanzi RE, Solomon F (1992) Identification of a mouse brain cDNA that encodes a protein related to the Alzheimer disease-associated amyloid β protein precursor. Proc Natl Acad Sci USA 89: 10785–10762

Wasco W, Gurubhagavatula S, d. Paradis M, Romano DM, S.S. S, Hyman BT, Neve RL, Tanzi RE (1993) Isolation and characterization of the human APLP2 gene encoding a homologue of the Alzheimer's associated amyloid β protein precursor. Nature Genet 5: 95–100

Weidemann A, Konig G, Bunke D, Fischer P, Salbaum JM, Masters CL, Beyreuther K (1989) Identification, biogenesis, and localization of precursors of Alzheimer's disease A4 amyloid protein. Cell 57: 115–126

Wolfe MS, Xia W, Ostaszewski BL, Diehl TS, Kimberly WT, Selkoe DJ (1999) Two transmembrane aspartates in presenilin-1 required for presenilin endoproteolysis and gamma-secretase activity. Nature 398: 513–517

Yamatsuji T, Matsui T, Okamoto T, Komatsuzaki K, Takeda S, Fukumoto H, Iwatsubo T, Suzuki N, Asami-Odaka A, Ireland S, Kinane TB, Giambarella U, Nishimoto I (1996) G protein-mediated neuronal DNA fragmentation induced by familial Alzheimer's disease-associated mutants of APP. Science 272: 1349–1352

Ye Y, Lukinova N, Fortini ME (1999) Neurogenic phenotypes and altered Notch processing in Drosophila Presenilin mutants. Nature 398: 525–529

Zheng H, Jiang M, Trumbauer ME, Sirinathsinghji DJ, Hopkins R, Smith DW, Heavens RP, Dawson GR, Boyce S, Conner MW (1995) β Amyloid precursor protein-deficient mice show reactive gliosis and decreased locomotor activity. Cell 81: 525–531

Zito K, Parnas D, Fetter RD, Isacoff EY, Goodman CS (1999) Watching a synapse grow: noninvasive confocal imaging of synaptic growth in Drosophila. Neuron 22: 719–729

A Gain of Function of the Huntington's Disease and Amyotrophic Lateral Sclerosis-Associated Genetic Mutations May Be a Loss of Bioenergetics

M. F. Beal

Summary

There is accumulating evidence for bioenergetic defects that may be involved in the pathogenesis of neurodegenerative diseases. In Huntington's disease (HD), the genetic defect is a CAG repeat expansion in a gene that encodes the protein huntingtin, whose function is unknown. Several lines of evidence have demonstrated that the HD mutation is associated with abnormalities in bioenergetics, both in lymphoblasts of patients as well as in postmortem brain material and in living patients, as assessed by MRI spectroscopy. Furthermore, recent studies in a transgenic mouse model of HD have shown that there are marked decreases in N-acetylaspartate and increases in glutamine consistent with a bioenergetic defect. In familial amyotrophic lateral sclerosis (FALS), there are point mutations in the enzyme copper/zinc superoxide dismutase (SOD1). Transgenic mice that overexpress SOD1 with FALS-associated mutations show prominent vacuolization of mitochondria, a finding that correlates with cell loss in the spinal cord as well as impaired motor function. Both the HD and FALS mutations appear to result in a gain of function. A consequence of this gain of function may be a deficit in bioenergetics. If this is the case, then several therapeutic strategies may be useful. We found that oral administration of the mitochondrial cofactor coenzyme Q_{10} or of the creatine kinase substrate creatine can significantly increase survival in transgenic mouse models of both ALS and HD. These may, therefore, be novel approaches for the treatment of these illnesses.

Introduction

Evidence implicating mitochondria in both necrotic and apoptotic cell death is rapidly accumulating (Nicotera and Lipton 1999). These conditions are distinct forms of cell death; however, they may either coexist or be sequential events, depending on the severity of the initiating insult. Cellular energy reserves appear to have an important role in these two forms of cell death with apoptosis favored under conditions in which ATP levels are preserved. Excitotoxic cell death, which is usually associated with glutamate toxicity, is associated with a prominent and persistent depolarization of the mitochondrial membrane potential followed by a depletion of ATP that results in necrosis (Schinder et al. 1996; White and

Research and Perspectives in Alzheimer's Diseases
Beyreuther/Christen/Masters (Eds.)
Neurodegenerative Disorders
© Springer-Verlag Berlin Heidelberg 2001

Reynolds 1996). Mitochondria are essential in controlling specific apoptosis pathways (Green and Reed 1998). They are responsible for the release of caspase activators such as cytochrome c, caspase-9, and apoptosis-inducing factor. Oversynthesis of the proapoptotic protein BAX triggers cytochrome c efflux from mitochondria (Desagher et al. 1999).

The role of apoptotic versus necrotic cell death in amyotrophic lateral sclerosis (ALS) and Huntington's disease (HD) is as yet controversial. There is evidence of increased DNA strand breaks, as detected by TUNEL staining in postmortem tissue from patients with HD and in patients with ALS (Dragunow et al. 1995; Martin 1999; Portera-Cailliau et al. 1995; Thomas et al. 1995). In HD the extent of TUNEL labeling correlates with CAG repeat length in the gene mutation. There are also increases in several proteins that are associated with apoptotic cell death. The recent study by Martin (1999) was a careful morphologic study that suggested that several of the features of apoptotic cell death were present in postmortem ALS tissue. These features include chromatin condensation in the nucleus.

A consequence of mitochondrial dysfunction is increased generation of free radicals and oxidative damage that are strongly implicated in the pathogenesis of neurodegenerative diseases. Mitochondria are the most important physiologic source of O_2^- in animal cells and are estimated to produce 2–3 nmols of O_2^-/min per milligram per mg/protein (Boveris and Chance 1973). If mitochondrial dysfunction plays a role in both ALS and HD, one would expect evidence of increased oxidative damage. This appears to be the case, as discussed below.

There is strong evidence that the HD mutation leads to a gain of function. It has been demonstrated that small deletions of the gene do not lead to the HD phenotype. Furthermore, a transgenic mouse with a knockout of the gene does not result in the disease phenotype (White et al. 1997). There is increasing evidence that the huntingtin protein is cleaved and translocated into the nucleus (Wheeler et al. 2000). This appears to be important for the eventual cell death. The development of intranuclear inclusions, however, does not appear to be critical for cell death (Klement et al. 1998; Saudou et al. 1998). Recent work in the transgenic mouse model of spinocerebellar ataxia-1 has demonstrated that a downregulation in gene expression is the earliest detectable biochemical event (Lin et al. 2000). Downregulation of three genes can be detected as early as three to four weeks of age (Lin et al. 2000). In contrast, the disease phenotype does not develop until 15 weeks of age. Downregulation of gene expression precedes the development of any phenotypic signs of ataxia or the development of intranuclear inclusions or cell loss. It is therefore plausible that proteins with CAG repeats maybe working by binding to transcription factors, which then affect gene regulation of multiple genes (Huang and Lee 1998; Kazantsev et al. 1999). One of the genes that is downregulated in the spinocerebellar ataxia-1 model is responsible for glutamate reuptake (Lin et al. 2000). A downregulation could lead to increased glutamate levels in the extracellular space, which could contribute to excitotoxic cell death.

There is substantial evidence that the HD mutation is associated with impaired energy metabolism. Consistent with this hypothesis, it has been found that lactate is elevated in the occipital cortex and basal ganglia in patients with HD and that there is a reduced phosphocreatine to inorganic phosphotate ratio in resting muscle of HD patients (Jenkins et al. 1993; Koroshetz et al. 1997). There are also decreases in N-acetylaspartate within the basal ganglia which correlate with the size of the CAG repeat expansion (Jenkins et al. 1998). Mitochondrial toxins, such as 3-nitropropionic acid (3-NP) produce selective damage to the basal ganglia, which closely replicates the pathology in HD (Beal et al. 1993; Brouillet and Hantraye 1995). 3-NP acid administration to both rats and primates results in a selective degeneration of spiny projection neurons within the striatum, a preservation of striatal afferents, a preservation of NADPH diaphorase interneurons, and spiny neuron dendritic abnormalities that are characteristically observed in HD postmortem tissue. Furthermore, these lesions in primates are associated with an early onset of choreiform movements followed by the onset of dystonia (Brouillet et al. 1995). They are also associated with a frontal type cognitive deficit in the absence of any cortical pathology (Palfi et al. 1996). The lesions are markedly age dependent. MRI spectroscopy shows that they are associated with focal increases in lactate confined to the basal ganglia. The increases in lactate in the occipital cortex of HD patients are significantly attenuated by administration of coenzyme Q_{10} (Koroshetz et al. 1997). This agent has also been shown to attenuate lesions produced by the mitochondrial toxins malonate and 3-NP (Beal et al. 1994). It also significantly reduces malonate-induced increases in striatal lactate concentrations. Recent studies have shown that reductions in both N-acetylaspartate and creatine-phosphocreatine concentrations in HD basal ganglia correlate with both clinical disability and CAG repeat expansions (Sanchez-Pernaute et al. 1999).

A number of biochemical studies in HD postmortem tissue show that there is a significant decrease in complex II–III activity in the caudate and putamen, with a smaller decrease in complex IV activity (Browne et al. 1997; Gu et al. 1996). The finding of a complex II–III defect in HD basal ganglia is of interest since inherited defects in complex II are associated with basal ganglia degeneration (Bourgeron et al. 1995). Furthermore, both malonate and 3-NP, which produce striatal degeneration that mimics that which occurs in HD, are selective complex II inhibitors. Recent studies in lymphoblast mitochondria in patients with HD show that they are abnormally susceptible to depolarization induced by low doses of cyanide (Sawa et al. 1999). The lymphoblast mitochondria depolarize significantly more readily than the control lymphoblasts. This degree of depolarization significantly correlated with CAG repeats length. The lymphocytes were also more susceptible to staurosporin-induced apoptotic cell death. The depolarization could be partially blocked by cyclosporin A, suggesting that it involved the permeability transition pore.

The mitochondrial permeability transition pore (PTP) may play a critical role in both necrotic and apoptotic cell death. Activation of the PTP increases the inner mitochondrial membrane permeability to solids with a mass up to 1.5 kDa. Proposed components of the PTP include the inner mitochondrial membrane adenine nucleotide transporter that interacts with cyclophilin D, and the voltage-dependent anion channel of the outer membrane (Bernardi et al. 1998). Mitochondrial creatine kinase, which is located between the inner and outer mitochondrial membranes, may also play a role in regulating the activation of the PTP. The opening of the channel is favored by elevated calcium and oxidizing agents, whereas closure is favored by low pH and adenine nucleotides such as ATP. Cyclosporin A acts as an excellent blocker of the channel by preventing an interaction of cyclophilin with the adenine nucleotide transporter.

Previous ultrastructural studies of cortical biopsies obtained from patients with both juvenile- and adult-onset HD show abnormal mitochondria (Goebel et al. 1978). Furthermore, there is progressive weight loss in HD patients despite high caloric intake (Obrien et al. 1990).

A breakthrough in HD research was the development of transgenic mouse models. Transgenic mice expressing exon-1 of the human HD gene with an expanded CAG repeat develop a progressive neurologic disorder (Mangiarini et al. 1996). These mice (line R6/2) have CAG repeat lengths of 141–157, under the control of the human HD promoter. At approximately six weeks of age the R6/2 mice show loss of brain and body weight, and at nine to eleven weeks they develop an irregular gait, abrupt shuttering, stereotypic movements, resting tremors, and epileptic seizures. The mice show an early decrease of several neurotransmitter receptors (Cha et al. 1998). The brains of the R6/2 mice appear normal in most respects; however, neuronal intranuclear inclusions that are immunopositive for huntingtin and ubiquitin are detected in the striatum at four and one half weeks of age (Davies et al. 1997). This finding is consistent with the finding of neuropil, cytoplasmic and nuclear inclusions in human HD (DiFiglia et al. 1997).

Using magnetic resonance spectroscopy we found a profound decrease in a N-acetylaspartate (NAA) concentrations in the R6/2 mice by six to seven weeks of age, yet no cell loss (Jenkins et al. 2000). Since NAA is synthesized within mitochondria, this may well reflect impaired mitochondrial function (Bates et al. 1996). We also found a significant increase in glutamine concentrations. Glutamine is synthesized by glutamine synthetase, which is a mitochondrial enzyme. The conversion of glutamine to glutamate and ammonia requires glutaminase, which is found within mitochondria of neurons. Impaired mitochondrial function may therefore be a cause of the significant increases in glutamine that we have observed.

There is also substantial evidence for energy defects in ALS. This is true in both sporadic ALS (SALS) and familial ALS (FALS) associated with point mutations in the enzyme copper/zinc SOD (Rosen et al. 1993). In patients with SALS. There are mitochondrial abnormalities and liver biopsies as well as anterior horn cells (Masui et al. 1985; Nakano et al. 1987; Sasaki et al. 1990). Recent studies of skeletal muscle biopsies of individuals with SALS show impairment in mitochondrial function (Wiedemann et al. 1998). There is a 50 % reduction in the specific activity of complex I as compared with age-matched controls and patients with spinal muscular atrophy. In addition, functional imaging of mitochondria using the ratios of NADPH and flavoprotein autofluorescence of permeabilized muscle fibers shows defective mitochondria in the single fiber level. There also appear to be increased mitochondrial DNA deletions and a decrease in mitochondrial DNA content (Vielhaber et al. 1999). Muscle biopsies of individuals with SALS also show increased mitochondrial volume and calcium levels (Siklos et al. 1996). A recent study looked at succinate dehydrogenase and cytochrome oxidase histochemical staining in the anterior horn cells of patients with sporadic ALS (Borthwick et al. 1999). A decrease in cytochrome oxidase activity is characteristic of mitochondrial defects. Succinate dehydrogenase is encoded on the nuclear genome. The authors demonstrated that patients with known mitochondrial DNA deletions showed a decrease in cytochrome oxidase in the anterior horn cells, and identical effects were seen in sporadic ALS patients. Prior studies showed that peripheral blood lymphocytes from individuals with ALS showed increased cytosolic calcium and impaired responses to uncouplers of oxidative phosphorylation (Curti et al. 1996).

A novel technique used to determine whether there are mitochondrial DNA defects associated with neurodegenerative diseases is to make cybrid cell lines (King and Attardi, 1989). These are made by fusing patients' platelets that contain only mitochondria with cell lines that are deficient in mitochondria. One therefore places the mitochondria into cell lines that have a different nuclear background. If a mitochondrial defect persists in these cell lines, it suggests that it is encoded on the mitochondrial DNA. Recent studies of ALS cybrids showed a significant decrease in complex I activity as well as trends toward reduced complex III–IV activities and increases in free radical scavenging enzymes (Swerdlow et al. 1998). An out of frame mutation of mitochondrial DNA encoded subunit 1 of cytochrome c oxidase was reported in an individual with motor neuron disease (Comi et al. 1998). There is substantial evidence for increased oxidative damage in individuals with SALS. There are reports of increased 8-hydroxy-2-deoxyguanosine in both motor cortex and spinal cord (Ferrante et al. 1997a). There are also increases in protein carbonyl groups in both spinal cord and motor cortex (Bowling et al. 1993; Shaw et al. 1995). There is an increase in 4-hydroxynonenal in the spinal cords of ALS patients (Pedersen et al. 1998). We found significant increases in 3-nitrotyrosine (3-NT) levels in the spinal cords of patients with both SALS and FALS associated with SOD1 mutations (Beal et al. 1997b). There are also reports of significant increases in 4-hydroxynonenal concentrations and 3-NT concentrations in spinal fluid of SALS patients (Smith et al. 1998; Tohgi et al. 1999). Our recent studies demonstrated significant increases in 8-hydroxy-2-deoxyguanosine concentrations in urine, plasma and CSF of ALS patients.

A major breakthrough in the study of ALS was the finding of point mutations in the enzyme copper/zinc SOD by Rosen and colleagues in 1993. This finding led to a number of hypotheses concerning the means by which these mutations caused cell death. It was demonstrated early on that the mutations result in a gain of function. This was suggested by the observation that a complete knockout of copper/zinc SOD did not result in a motor neuron phenotype (Reaume et al. 1996). In contrast, overexpression of mutations in SOD1 resulted in progressive degeneration of anterior horn cells as well as progressive paralysis in transgenic mice (Gurney et al. 1994). There have been two major hypotheses concerning the gain of function of the mutant protein. It was initially hypothesized that the mutations resulted in an abnormal catalytic activity of the active site copper. It was suggested that there could be increased reactivity with either hydrogen peroxide or peroxynitrite (Beckman and Crow 1993; Wiedau-Pazos et al. 1996; Yim et al. 1996). More recent work suggests that zinc-deficient enzyme can directly generate peroxynitrite in the active site (Estevez et al. 1999). It has been demonstrated that the mutations in SOD1 result in a decreased affinity of zinc for the enzyme (Crow et al. 1997; Lyons et al. 1996). This destabilizes the backbone of the enzyme, which may lead to an increased accessibility of superoxide and nitric oxide to the active site copper. It is also possible that the active site copper may be reduced by cellular reductants, such as ascorbate or glutathione, and may then be able to react with oxygen to generate superoxide in the active site, which could then directly react with nitric oxide to generate peroxynitrite (Estevez et al. 1999). This process has been demonstrated in cultured motor neurons.

The other major hypothesis is that the mutations may lead to increased protein aggregation. Precisely how this would lead to cell death is unclear. Increased protein aggregation is, however, a well-known feature of a number of neurodegenerative diseases (Durham et al. 1997; Ross 1997), including HD as discussed above (DiFiglia et al. 1997). An argument against an increased generation of free radicals was the observation that crossing transgenic ALS mice either with mice with a knockout of SOD or with mice overexpressing wild-type SOD did not modify the phenotype (Bruijn et al. 1998). In contrast, we have found that crossing the transgenic ALS mice with mice with a 50 % reduction in manganese SOD significantly accelerates the phenotype and the cell loss and mitochondrial vacuolization that are known to be characteristic features of mice with the G93A SOD1 mutation (Andreassen et al. 2000).

Expression of SOD1 with the G93A mutation in neuroblastoma cells results in a loss of mitochondrial membrane potential as well as an elevation of cytosolic calcium concentrations (Carri et al. 1997). Neuropathological studies of two lines of the transgenic ALS mice with SOD1 mutations have demonstrated that vacuolization of mitochondria is an early and prominent pathological feature (Gurney et al. 1994; Wong et al. 1995). The increase in mitochondrial vacuolization immediately precedes a rapid phase of motor weakness, loss of motor neurons and loss of axons (Kong and Xu 1997). At the electron microscopic level there appears to be a splitting of the inner and outer mitochondrial membrane (Wong et al. 1995). The vacuolated mitochondria are also associated with glial processes. A number of studies have demonstrated evidence of increased oxidative damage in G93A mice. There is increased immunostaining for malondialdehyde, 8-hydroxy-2-deoxyguanosine and 3-NT in the motor neurons (Ferrante et al. 1997b). We demonstrated significantly increased levels of 3-NT in two different transgenic mouse lines associated with ALS (Ferrante et al. 1997b). There is also evidence of increased protein carbonyl groups, malondialdehyde and 8-hydroxy-2-deoxyguanosine, as determined biochemically in the spinal cords of the G93A transgenic ALS mice (Andrus et al. 1998; Liu et al. 1998). The evidence therefore implicates mitochondrial dysfunction in both SALS and FALS associated with SOD1 mutations.

If bioenergetic defects are associated with both ALS and HD, then agents that might improve bionergetics could prove useful. There has been considerable interest in the use of coenzyme Q_{10} for treatment of mitochondrial disorders. This compound has been reported to improve ATP generation in vitro and it serves as an important antioxidant in both mitochondrial and lipid membranes. Oral administration of coenzyme Q_{10} significantly attenuates dopamine depletion produced by MPTP (Beal et al. 1997a). It also significantly atttenuates lesions produced by intrastriatal administration of malonate and 3-NP acid (Beal et al. 1994; Matthews et al. 1998b), and significantly extends survival in a transgenic mouse model of ALS (Matthews et al. 1998b). We have recently found that oral administration of coenzyme Q_{10} can significantly increase survival of the R6/2 transgenic mouse model of HD.

Another potential therapeutic strategy for diseases associated with mitochondrial dysfunction is to use creatine to increase brain energy stores and thereby compensate for an energetic defect. Creatine kinase and its substrates creatine and phosphocreatine constitute an intricate cellular energy buffering and transport system connecting sites of energy production (mitochondria) with sites of energy consumption (Hemmer and Wallimann 1993). Creatine administration increases brain concentrations of phosphocreatine and inhibits activation of the mitochondrial PTP (Matthews et al. 1998a; O'Gorman et al. 1996). There is evidence for a direct functional coupling of creatine kinase with sodium potassium ATPase, neurotransmitter release, maintenance of memory potentials and restoration of ion gradients after depolarization (Dunant et al. 1988; Hemmer and Wallimann 1993). Creatine kinase flux correlates with brain activity, as measured by the EEG, as well as with amounts of 2-deoxyglucose uptake in the brain (Corbett and Laptook 1994; Sauter and Rudin 1993). Creatine kinase also appears to be coupled directly or indirectly to energetic processes required for calcium homeostasis (Deshpande et al. 1997; Steeghs et al. 1997). Phosphocreatine can also directly stimulate synaptic glutamate uptake and thereby reduce extracellular glutamate (Xu et al. 1996). The effects of creatine on the PT appear to be through its ability to stabilize mitochondrial creatine kinase in an octomeric form that inhibits activation (O'Gorman et al. 1997). Creatine administration also inhibits peroxynitrite-induced modification and inactivation of creatine kinase.

Prior studies have shown that creatine reduces oxidative damage to hippocampal slices in vitro (Carter et al. 1995). We found that creatine administration exerts neuroprotective effects against both the mitochondrial toxins malonate and 3-NP (Matthews et al. 1998b). It also significantly attenuated increases in lactate produced by systemic administration of 3-NP. Creatine administration also attenuates MPTP-induced dopamine depletions and substantia nigra neuronal loss (Matthews et al. 1999), and increases survival and improves motor performance in a transgenic mouse model of ALS, resulting in marked neuroprotective effects against the loss of anterior horn motor neurons and substantia nigra dopaminergic neurons (Klivenyi et al. 1999). We recently examined whether creatine could exert neuroprotective effects in a transgenic mouse model of HD (R6/2). We found that creatine dose-dependently improved survival in these mice (Ferrante et al. 2000). Creatine administration also resulted in improved rotarod performance and reduced weight loss in the R6/2 mice. A dose of 2% creatine was most efficacious, with 1% creatine and 3% creatine being less efficacious. Administration of creatine also significantly delayed the onset of diabetes, which occurs in the R6/2 mice. Interestingly, administration of creatine significantly delayed the development of neuronal shrinkage and gross atrophy of the brain, and significantly slowed the early decreases in N-acetylaspartate concentrations. Lastly the administration of creatine slowed the development of nuclear intra-neuronal inclusions.

Conclusions

The gain of function in both HD and ALS associated with dominant mutations appears to lead to a loss of function. We believe that this loss of function includes decreases in bioenergetic function as described above. If mitochondrial dysfunction does have a causative role in disease pathogenesis, then a number of therapeutic targets are implicated, including the PTP, cytochrome c released from mitochondria and free radical scavengers. It might also be possible to buffer energy levels in the brain using coenzyme Q_{10} or creatine as novel therapeutic strategies.

Acknowledgments

This work was supported by grants from the NIH, the Department of Defense, ALS Association, the Huntington's Disease Society of America and the Hereditary Disease Foundation. The secretarial assistance of Sharon Melanson is gratefully acknowledged.

References

Andreassen OA, Ferrante RJ, Klivenyi P, Klein AM, Shinobu LA, Epstein CJ, Beal MF (2000) Partial deficiency of manganese superoxide dismutase exacerbates a transgenic mouse model of amyotrophic lateral sclerosis. Ann Neurod 47: 447–455

Andrus PK, Fleck TJ, Gurney ME, Hall ED (1998) Protein oxidative damage in a transgenic mouse model of familial amyotrophic lateral sclerosis. J Neurochem 71: 2041–2048

Bates TE, Strangward M, Keelan J, Davey GP, Munro PMG, Clark JB (1996) Inhibition of N-acetylaspartate production: implications for ^{1}H MRS studies in vivo. NeuroReport 7: 1397–1400

Beal MF, Brouillet E, Jenkins BG, Ferrante RJ, Kowall NW, Miller JM, Storey E, Srivastava R, Rosen BR, Hyman BT (1993) Neurochemical and histologic characterization of striatal excitotoxic lesions produced by the mitochondrial toxin 3-nitropropionic acid. J Neurosci 13: 4181–4192

Beal MF, Henshaw R, Jenkins BG, Rosen BR, Schulz JB (1994) CoenzymeQ$_{10}$ and nicotinamide block striatal lesions produced by the mitochondrial toxin malonate. Ann Neurol 36: 882–888

Beal MF, Matthews R, Tieleman A, Schults CW (1997a) Coenzyme Q$_{10}$ attenuates the MPTP induced loss of striatal dopamine and dopaminergic axons in aged mice. Brain Res 783: 109–114

Beal MF, Shinobu LA, Schulz JB, Matthews RT, Thomas CE, Kowall NW, Gurney ME, Ferrante RJ (1997b) Increased 3-nitrotyrosine and oxidative damage in mice with a human Cu, Zn superoxide dismutase. Neurol Abst 48: A149

Beckman JS, Crow JP (1993) Pathological implications of nitric oxide superoxide and peroxynitrite formation. Biochem Soc Trans 21: 330–334

Bernardi P, Basso E, Colonna R, Costantini P, Di Lisa F, Eriksson O, Fontaine E, Forte M, Ichas F, Massari S, Nicolli A, Petronilli V, Scorrano L (1998) Perspectives of the mitochondrial permeability transition. Biochim Biophys Acta 1365: 200–206

Borthwick GM, Johnson MA, Ince PG, Shaw PJ, Turnbull DM (1999) Mitochondrial enzyme activity in amyotrophic lateral sclerosis: implications for the role of mitochondria in neuronal cell death. Ann Neurol 46: 787–790

Bourgeron T, Rustin P, Chretien D, Birch-Machin M, Bourgeois M, Viegas-Pequignot E, Munnich A, Rotig A (1995) Mutation of a nuclear succinate dehydrogenase gene results in mitochondrial respiratory chain deficiency. Nature Genet 11: 144–149

Boveris A, Chance B (1973) The mitochondrial generation of hydrogen peroxide. Biochem J 134: 707–716

Bowling AC, Schulz JB, Brown Jr RH, Beal MF (1993) Superoxide dismutase activity, oxidative damage, and mitochondrial energy metabolism in familial and sporadic amyotrophic lateral sclerosis. J Neurochem 61: 2322–2325

Brouillet E, Hantraye P (1995) Effects of chronic MPTP and 3-nitropropionic acid in nonhuman primates. Curr Opin Neurol 8: 469–473

Brouillet E, Hantraye P, Ferrante RJ, Dolan R, Leroy-Willig A, Kowall NW, Beal MF (1995) Chronic mitochondrial energy impairment produces selective striatal degeneration and abnormal choreiform movements in primates. Proc Natl Acad Sci USA 92: 7105–7109

Browne SE, Bowling AC, MacGarvey U, Baik MJ, Berger SC, Muqit MMK, Bird ED, Beal MF (1997) Oxidative damage and metabolic dysfunction in Huntington's disease: selective vulnerability of the basal ganglia. Ann Neurol 41: 646–653

Bruijn LI, Houseweart MK, Kato S, Anderson KL, Anderson SD Ohama E, Reaume AG, Scott RW, Cleveland DW (1998) Aggregation and motor neuron toxicity of an ALS-linked SOD1 mutant independent from wild-type SOD1. Science 281: 1851–1854

Carri MT, Ferri A, Battistoni A, Famhy L, Gabbianelli R, Poccia F, Rotilio G (1997) Expression of a Cu, Zn superoxide dismutase typical of familial amyotrophic lateral sclerosis induces mitochondrial alteration and increase of cytosolic Ca^{2+} concentration in transfected neuroblastoma SH-SY5Y cells. FEBS Lett 414: 365–368

Carter AJ, Muller RE, Pschorn U, Stransky W (1995) Preincubation with creatine enhances levels of creatine phosphate and prevents anoxic damage in rat hippocampal slices. J Neurochem 64: 2691–2699

Cha J-HJ, Kosinski CM, Kerner JA, Alsdorf SA, Mangiarini L, Davies SW, Penney JB, Bates GP, Young AB (1998) Altered brain neurotransmitter receptors in transgenic mice expressing a portion of an abnormal human Huntington disease gene. Proc Natl Acad Sci USA 95: 6480–6485

Comi GP, Bordoni A, Salani S, Fransceschina L, Sciacco M, Prelle A, Fortunato F, Zeviani M, Napoli L, Bresolin N, Moggio M, Ausenda CD, Taanman J-W, Scarlato G (1998) Cytochrome c oxidase subunit I microdeletion in a patient with motor neuron disease. Ann Neurol 43: 110–116

Corbett RJT, Laptook AR (1994) Age-related changes in swine brain creatine kinase-catalyzed ^{31}P exchange measured in vivo using ^{31}P NMR magnetization transfer. J Cereb Blood Flow Metab 14: 1070–1077

Crow JP, Ye YZ, Strong M, Kirk M, Barnes S, Beckman JS (1997) Superoxide dismutase catalyzes nitration of tyrosines by peroxynitrite in the rod and head domains of neurofilament-L. J Neurochem 69: 1945–1953

Curti D, Malaspina A, Facchetti G, Camana C, Mazzini L, Tosca P, Zerbi F, Ceroni M (1996) Amyotrophic lateral sclerosis: Oxidative energy metabolism and calcium homeostasis in peripheral blood lymphocytes. Neurology 47: 1060–1064

Davies SW, Turmaine M. Cozens BA, DiFiglia M, Sharp AH, Ross CA, Scherzinger E, Wanker EE, Mangiari L, Bates GP (1997) Formation of neuronal intranuclear inclusions underlies the neurological dysfunction in mice transgenic for the HD mutation. Cell 90: 537–548

Desagher S, Osen-Sand A, Nichols A, Eskes R, Montessuit S, Lauper S, Maundrell K, Antonsson B, Martinou JC (1999) Bid-induced conformational change of Bax is responsible for mitochondrial cytochrome c release during apoptosis. J Cell Biol 144: 891–901

Deshpande SB, Fukuda A, Nishino H (1997) 3-Nitropropionic acid increases the intracellular Ca^{2+} in cultured astrocytes by reverse operation of the Na^{+}-Ca^{2+} exchanger. Exp Neurol 145: 38–45

DiFiglia M, Sapp E, Chase KO, Davies SW, Bates GP, Vonsattel JP, Aronin N (1997) Aggregation of huntingtin in neuronal intranuclear inclusions and dystrophic neurites in brain. Science 277: 1990–1993

Dragunow M, Faull RLM, Lawlor P, Beilharz EJ, Singleton K, Walker EB, Mee E (1995) In situ evidence for DNA fragmentation in Huntington's disease striatum and Alzheimer's disease temporal lobes. Neuroreport 6: 1053–1057

Dunant Y, Loctin F, Marsal J, Muller D, Parducz A, Rabasseda X (1988) Energy metabolism and quantal acetylcholine release: effects of botulinum toxin, 1-fluoro-2,4-dinitrobenzene, and diamide in the Torpedo electric organ. J Neurochem 50: 431–439

Durham HD, Roy J, Dong L, Figlewicz DA (1997) Aggregation of mutant Cu/ZN superoxide dismutase proteins in a culture model of ALS. J Neuropath Exp Neurol 56: 523–530

Estevez AG, Crow JP, Sampson JB, Reiter C, Zhuang Y, Richardson GJ, Tarpey MM, Barbeito L, Beckman JS (1999) Induction of nitric oxide-dependent apoptosis in motor neurons by zinc-deficient superoxide dismutase. Science 286: 2498–2500

Ferrante RJ, Browne SE, Shinobu LA, Bowling AC, Baik MJ, MacGarvey U, Kowall NW, Brown Jr, RH, Beal MF (1997a) Evidence of increased oxidative damage in both sporadic and familial amyotrophic lateral sclerosis. J Neurochem 69: 2064–2074

Ferrante RJ, Shinobu LA, Schulz JB, Matthews RT, Thomas CE, Kowall NW, Gurney ME, Beal MF (1997b) Increased 3-nitrotyrosine and oxidative damage in mice with a human Cu, Zn superoxide dismutase mutation. Ann Neurol 42: 326–334

Ferrante RJ, Andreassen OA, Jenkins BG, Dedeoglu A, Kuemmerle S, Kubilus JK, Kaddurah-Daouk R, Hersch SM, Beal MF (2000) Neuroprotective effects of creatine in a transgenic mouse model of Huntington's disease. J Neurosci 20: 4384–4397

Goebel HH, Heipertz R, Scholz W, Iqbal K, Tellez-Nagel I (1978) Juvenile Huntington chorea: clinical, ultrastructural, and biochemical studies. Neurology 28: 23–31

Green DR, Reed JC (1998) Mitochondria and apoptosis. Science 281: 1309–1312

Gu M, Gash MT, Mann VM, Javoy-Agid F, Cooper JM, Schapira AHV (1996) Mitochondrial defect in Huntington's disease caudate nucleus. Ann Neurol 39: 385–389

Gurney ME, Pu H, Chiu AY, Dal Canto MC, Polchow CY, Alexander DD, Caliendo J, Hentati A, Kwon YW, Deng H-X, Chen W, Zhai P, Sufit RL, Siddique T (1994) Motor neuron degeneration in mice that express a human Cu, Zn superoxide dismutase mutation. Science 264: 1772–1775

Hemmer W, Wallimann T (1993) Functional aspects of creatine kinase in brain. Dev Neurosci 15: 249–260

Huang H-C, Lee EHY (1998) MPTP produces differential oxidative stress and antioxidative responses in the nigrostriatal and mesolimbic dopaminergic pathways. Free Radic Biol Med 24: 76–84

Jenkins BG, Koroshetz WJ, Beal MF, Rosen BR (1993) Evidence for impairment of energy metabolism in vivo in Huntington's disease using localized 1H NMR spectroscopy. Neurology 43: 2689–2695

Jenkins BG, Rosas HD, Chen YC, Makabe T, Myers R, MacDonald M, Rosen BR, Beal MF, Koroshetz WJ (1998) 1H NMR spectroscopy studies of Huntington's disease: correlations with CAG repeat numbers. Neurology 50: 1357–1365

Jenkins BG, Klivenyi P, Kustermann E, Andreassen OA, Ferrante RJ, Rosen BR, Beal MF (2000) Nonlinear decrease over time in n-acetylaspartate levels in the absence of neuronal loss and increases in glutamine and glucose in transgenic Huntington's disease mice. J Neurochem 74: 2108–2119

Kazantsev A, Preisinger E, Dranovsky A, Goldgaber D, Housman D (1999) Insoluble detergent-resistant aggregates form between pathological and nonpathological lengths of polyglutamine in mammalian cells. Proc Natl Acad Sci USA 96: 11404–11409

King MP, Attardi G (1989) Human cells lacking mtDNA: repopulation with exogenous mitochondria by complementation. Science 246: 500–503

Klement IA, Skinner PJ, Kaytor MD, Yi H, Hersch SM, Clark HB, Zoghbi HY, Orr HT (1998) Ataxin-1 nuclear localization and aggregation: role in polyglutamine-induced disease in SCA1 transgenic mice. Cell 95: 41–53

Klivenyi P, Ferrante RJ, Matthews RT, Bogdanov MB, Klein AM, Andreassen OA, Mueller G, Wermer M, Kaddurah-Daouk R, Beal MF (1999) Neuroprotective effects of creatine in a transgenic animal model of amyotrophic lateral sclerosis. Nature Med 5: 347–350

Kong J-M, Xu Z-S (1997) Four stages of disease progression in an ALS mouse model. Soc Neurosci Abst 23: 1913

Koroshetz WJ, Jenkins BG, Rosen BR, Beal MF (1997) Energy metabolism defects in Huntington's disease and effects of coenzyme Q10. Ann Neurol 41: 160–165

Lin X, Antalffy B, Kang D, Orr HT, Zoghbi HY (2000) Polyglutamine expansion down-regulates specific neuronal genes before pathologic changes in SCA1. Nature Neurosci 3: 157–163

Liu R, Althaus JS, Ellerbrock BR, Becker DA, Gurney ME (1998) Enhanced oxygen radical production in a transgenic mouse model of familial amyotrophic lateral sclerosis. Ann Neurol 44: 763–770

Lyons TJ, Liu H, Goto JJ, Nersissian A, Roe JA, Graden JA, Cafe C, Ellerby LM, Bredesen DE, Butler-Gralla E, Selverstone-Valentine J (1996) Mutations in copper-zinc superoxide dismutase that cause amyotrophic lateral sclerosis alter the zinc binding site and the redox behavior of the protein. Proc Natl Acad Sci USA 93: 12240–12244

Mangiarini L, Sathasivam K, Seller M, Cozens B, Harper A, Hetherington C, Lawton M, Trottier Y, Lehrach H, Davies SW, Bates GP (1996) Exon 1 of the HD gene with an expanded CAG repeat is sufficient to cause a progressive neurological phenotype in transgenic mice. Cell 87: 493–506

Martin LJ (1999) Neuronal death in amyotrophic lateral sclerosis is apoptosis: possible contribution of a programmed cell death mechanism. J Neuropathol Exp Neurol 58: 459–471

Masui Y, Mozai T, Kakehi K (1985) Functional and morphometric study of the liver in motor neuron disease. J Neurol 232: 15–19

Matthews RT, Yang L, Jenkins BG, Ferrante RJ, Rosen BR, Kaddurah-Daouk R, Beal MF (1998a) Neuroprotective effects of creatine and cyclocreatine in animal models of Huntington's disease. J Neurosci 18: 156–163

Matthews RT, Yang S, Browne S, Baik M, Beal MF (1998b) Coenzyme Q_{10} administration increases brain mitochondrial concentrations and exerts neuroprotective effects. Proc Natl Acad Sci USA 95: 8892–8897

Matthews RT, Ferrante RJ, Klivenyi P, Yang L, Klein AM, Mueller G, Kaddurah-Daouk R, Beal MF (1999) Creatine and cyclocreatine attenuate MPTP neurotoxicity. Exp Neurol 157: 142–149

Nakano K, Hirayama K, Terao K (1987) Hepatic ultrastructural changes and liver dysfunction in amyotrophic lateral sclerosis. Arch Neurol 44: 103–106

Nicotera P, Lipton SA (1999) Excitotoxins in neuronal apoptosis and necrosis. J Cereb Blood Flow Metab 19: 583–591

O'Gorman E, Beutner G, Wallimann T, Brdiczka D (1996) Differential effects of creatine depletion on the regulation of enzyme activities and on creatine-stimulated mitochondrial respiration in skeletal muscle, heart, and brain. Biochim Biophys Acta 1276: 161–170

O'Gorman E, Beutner G, Dolder M, Koretsky AP, Brdiczka D, Wallimann T (1997) The role of creatine kinase inhibition of mitochondrial permeability transition. FEBS Lett 414: 253–257

Obrien CF, Miller C, Goldblatt D, Welle S, Forbes G, Lipinski B, Panzik J, Peck R, Plumb S, Oakes D, Kurlan R, Shoulson I (1990) Extraneural metabolism in early Huntington's disease. Ann Neurol 28: 300–301

Palfi S, Ferrante RJ, Brouillet E, Beal MF, Dolan R, Guyoi MC, Peschanski M, Hantraye P (1996) Chronic 3-nitropropionic acid treatment in baboons replicates the cognitive and motor deficits of Huntington's disease. J Neurosci 16: 3019–3025

Pedersen WA, Fu W, Keller JN, Markesbery WR, Appel S, Smith G, Kasarskis E, Mattson MP (1998) Protein modification by the lipid peroxidation product 4-hydroxynonenal in the spinal cords of amyotrophic lateral sclerosis patients. Ann Neurol 44: 819–824

Portera-Cailliau C, Hedreen JC, Price DL, Koliatsos VE (1995) Evidence for apoptotic cell death in Huntington's disease and excitotoxic animal models. J Neurosci 15: 3775–3787

Reaume AG, Elliott JL, Hoffman EK, Kowall NW, Ferrante RJ, Siwek DE, Wilcox HM, Flood DG, Beal MF, Brown Jr, RH, Scott RW, Snider WD (1996) Motor neurons in Cu/Zn superoxide dismutase-deficient mice develop normally but exhibit enhanced cell death after axonal injury. Nature Genet 13: 43–47

Rosen DR, Siddique T, Patterson D, Figiewicz DA, Sapp P, Hentati A, Donaldson D, Goto J, O'Regan JP, Deng H-X, Rhmani Z, Krizus A, McKenna-Yasek D, Cayabyab A, Gaston SM, Berger R, Tanzi RE, Halperin JJ, Herzfeldt B, Van den Bergh R, Hung W-Y, Bird T, Deng G, Mulder DW, Smyth C, Laing NG, Soriano E, Pericak-Vance MA, Haines J, Rouleau GA, Gusella JS, Horvitz HR, Brown RH (1993) Mutations in Cu/Zn superoxide dismutase gene are associated with familial amyotrophic lateral sclerosis. Nature 362: 59–62

Ross CA (1997) Intranuclear neuronal inclusions: a common pathogenic mechanism for glutamine-repeat neurodegenerative diseases? Neuron 19: 1147–1150

Sanchez-Pernaute R, Garcia-Segura JM, del Barrio Alba A, Viano J, de Yebenes JG (1999) Clinical correlation of striatal 1H MRS changes in Huntington's disease. Neurology 53: 806–812

Sasaki S, Maruyama S, Yamane K, Sakuma H, Takeishi M (1990) Ultrastructure of swollen proximal axons of anterior horn neurons in motor neuron disease. J Neurol Sci 97: 233–240

Saudou F, Finkbeiner S, Devys D, Greenberg ME (1998) Huntingtin acts in the nucleus to induce apoptosis but death does not correlate with the formation of intranuclear inclusions. Cell 95: 55–66

Sauter A, Rudin M (1993) Determination of creatine kinase parameters in rat brain by NMR magnetization transfer: correlation with brain function. J Biol Chem 268: 13166–13171

Sawa A, Wiegand GW, Cooper J, Margolis RL, Sharp AH, Lawler JF, Jr., Greenamyre JT, Snyder SH, Ross CA (1999) Increased apoptosis of Huntington disease lymphoblasts associated with repeat length-dependent mitochondrial depolarization. Nature Med 5: 1194–1198

Schinder AF, Olson EC, Spitzer NC, Montal M (1996) Mitochondrial dysfunction is a primary event in glutamate neurotoxicity. J Neurosci 16: 6125–6133

Shaw PJ, Ince PG, Falkous G, Mantle D (1995) Oxidative damage to protein in sporadic motor neuron disease spinal cord. Ann Neurol 38: 691–695

Siklos L, Engelhardt J, Harati Y, Smith RG, Joo F, Appel SH (1996) Ultrastructural evidence for altered calcium in motor nerve terminals in amyotrophic lateral sclerosis. Ann Neurol 39: 203–219

Smith RG, Henry YK, Mattson MP, Appel SH (1998) Presence of 4-hydroxynonenal in cerebrospinal fluid of patients with sporadic amyotrophic lateral sclerosis. Ann Neurol 44: 696–699

Steeghs K, Benders A, Oerlemans F, de Haan A, Heerschap A, Ruitenbeek W, Jost C, van Deursen J, Perryman B, Pette D, Bruckwilder M, Koudijs J, Jap P, Veerkamp J, Wieringa B (1997) Altered Ca^{2+} responses in muscles with combined mitochondrial and cytosolic creatine kinase deficiencies. Cell 89: 93–103

Swerdlow RH, Parks JK, Cassarino DS, Trimmer PA, Miller SW, Maguire DJ, Sheehan JP, Maguire RS, Pattee G, Juel VC, Phillips LH, Tuttle JB, Bennett J, JP, Davis RE, Parker JWD (1998) Mitochondria in sporadic amyotrophic lateral sclerosis. Exp Neurol 153: 135–142

Thomas LB, Gates DJ, Richfield EK, O'Brien TF, Schweitzer JB, Steindler DA (1995) DNA end labeling (TUNEL) in Huntington's disease and other neuropathological conditions. Exp Neurol 133: 265–272

Tohgi H, Abe T, Yamazaki K, Murata T, Ishizaki E, Isobe C (1999) Remarkable increase in cerebrospinal fluid 3-nitrotyrosine in patients with sporadic amyotrophic lateral sclerosis. Ann Neurol 46: 129–131

Vielhaber S, Winkler K, Kirches E, Kunz D, Buchner M, Feistner H, Elger CE, Ludolph AC, Riepe MW, Kunz WS (1999) Visualization of defective mitochondrial function in skeletal muscle fibers of patients with sporadic amyotrophic lateral sclerosis. J Neurol Sci 169: 133–139

Wheeler VC, White JK, Gutekunst CA, Vrbanac V, Weaver M, Li XJ, Li SH, Yi H, Vonsattel JP, Gusella JF, Hersch S, Auerbach W, Joyner AL, MacDonald ME (2000) Long glutamine tracts cause nuclear localization of a novel form of huntingtin in medium spiny striatal neurons in Hdh(Q92) and Hdh(Q111) knock-in mice. Human Mol Genet 9: 503–513

White JK, Auerbach W, Duyao MP, Vonsattel J-P, Gusella JE, Joyner AL, MacDonald ME (1997) Huntingtin is required for neurogenesis and is not impaired by the Huntington's disease CAG expansion. Nature Genet 17: 404–410

White RJ, Reynolds IJ (1996) Mitochondrial depolarization in glutamate-stimulated neurons: an early signal specific to excitotoxin exposure. J Neurosci 16: 5688–5697

Wiedau-Pazos M, Goto JJ, Rabizadeh S, Gralla EB, Roe JA, Lee MK, Valentine JS, Bredesen DE (1996) Altered reactivity of superoxide dismutase in familial amyotrophic lateral sclerosis. Science 271: 515–518

Wiedemann FR, Winkler K, Kuznetsov AV, Bartels C, Vielhaber S, Feistner H, Kunz WS (1998) Impairment of mitochondrial function in skeletal muscle of patients with amyotrophic lateral sclerosis. J Neurol Sci 156: 65–72

Wong PC, Pardo CA, Borchelt DR, Lee MK, Copeland NG, Jenkins NA, Sisodia SS, Cleveland DW, Price DL (1995) An adverse property of a familial ALS-linked SOD1 mutation causes motor neuron disease characterized by vacuolar degeneration of mitochondria. Neuron 14: 1105–1116

Xu CJ, Klunk WE, Kanfer JN, Xiong Q, Miller G, Pettegrew JW (1996) Phosphocreatine-dependent glutamate uptake by synaptic vesicles. J Biol Chem 271: 13435–13440

Yim MB, Kang JH, Yim HS, Kwak HS, Chock PB, Stadtman ER (1996) A gain-of-function of an amyotrophic lateral sclerosis-associated Cu, Zn-superoxide dismutase mutant: an enhancement of free radical formation due to a decrease in K_m for hydrogen peroxide. Proc Natl Acad Sci USA 93: 5709–5714

Subject Index

Printing (Computer to Film): Saladruck, Berlin
Binding: Stürtz AG, Würzburg